SpringerBriefs in Applied Sciences and Technology

Structural Mechanics

Series editor

Emmanuel E. Gdoutos, Department of Theoretical and Applied Mechanics, Democritus University of Thrace, Xanthi, Greece

More information about this series at http://www.springer.com/series/15039

Riadh Al-Mahaidi · Javad Hashemi
Robin Kalfat · Graeme Burnett
John Wilson

Multi-axis Substructure Testing System for Hybrid Simulation

 Springer

Riadh Al-Mahaidi
Smart Structures Laboratory
Swinburne University of Technology
Melbourne, VIC
Australia

Graeme Burnett
Smart Structures Laboratory
Swinburne University of Technology
Melbourne, VIC
Australia

Javad Hashemi
Smart Structures Laboratory
Swinburne University of Technology
Melbourne, VIC
Australia

John Wilson
Smart Structures Laboratory
Swinburne University of Technology
Melbourne, VIC
Australia

Robin Kalfat
Smart Structures Laboratory
Swinburne University of Technology
Melbourne, VIC
Australia

ISSN 2191-530X   ISSN 2191-5318   (electronic)
SpringerBriefs in Applied Sciences and Technology
ISSN 2520-8020   ISSN 2520-8039   (electronic)
Structural Mechanics
ISBN 978-981-10-5866-0   ISBN 978-981-10-5867-7   (eBook)
https://doi.org/10.1007/978-981-10-5867-7

Library of Congress Control Number: 2017951985

Printed on acid-free paper

This Springer imprint is published by Springer Nature
The registered company is Springer Nature Singapore Pte Ltd.
The registered company address is: 152 Beach Road, #21-01/04 Gateway East, Singapore 189721, Singapore

The original version of the book was revised:
For detailed information please see Erratum.
The erratum to the book is available at
https://doi.org/10.1007/978-981-10-5867-7_6

Acknowledgements

The authors gratefully acknowledge the contribution of the Australian Research Council (Grants LE110100052, DP140103350 and DP1096753) and 11 Australian partner universities for their assistance with the establishment of the 6-DOF hybrid testing facility. The authors would like to acknowledge the contribution of Ph.D. students Scott Menegon and Yassamin Al-Ogaidi and the personnel of the Smart Structures Laboratory at Swinburne University of Technology, including Michael Culton, Kia Rasekhi, Sanjeet Chandra and Kevin Nievaart.

Contents

List of Figures

List of Tables

Abstract

Hybrid simulation combines the flexibility and cost-effectiveness of computer simulation with the realism of experimental testing to provide a powerful tool for investigating the effects of extreme loads on large-scale structures. The key advantage is that only the critical components of a structure that are difficult to model numerically are sub-structured for testing in the laboratory, while the remainder of the structure with more predictable behavior is computer simulated using finite-element analysis software. The quality of the hybrid simulation relies to a great extent on the correct application of the interface boundary conditions between the numerical and the physical sub-domains. In addition, in order to evaluate the life-cycle capacity of a prototype structure, the loads must be applied on the hybrid model in the same manner as in the prototype, namely, gravity load first, followed by a sequence of service and/or extreme loads. To meet these objectives, a state-of-the-art loading system, referred to as the Multi-Axis Substructure Testing (MAST) system, has been designed, assembled and validated at Swinburne University of Technology, Melbourne, Australia, to expand the capabilities of hybrid testing to include three-dimensional responses of structures through mixed load/deformation control of six degrees-of-freedom (6-DOF) boundary conditions. This cutting-edge facility is unique in Australasia and is capable of serving the research community and practice, nationally and internationally. This report presents the design details of the MAST system, including the strong reaction wall/floor, the cruciform crosshead, servo-hydraulic actuators and the 6-DOF controller system and hybrid simulation architecture. In addition, to demonstrate the unique and powerful capabilities of the MAST system, specifically for collapse assessment of structures, the results of three experiments, including a quasi-static cyclic and two hybrid simulation test series, are presented.

Keywords Experimental techniques · Hybrid simulation · Large-scale testing · Multi-directional loading · Mixed mode control · Collapse assessment

Chapter 1
Introduction

Abstract This chapter discusses the role of experimental methods in earthquake engineering with a brief summary of advantages and challenges of the various test methods used in this field. The objectives and motivations of this research are discussed with an overview of the contents of this book.

Keywords Experimental techniques · Hybrid simulation · Structural dynamics · Earthquake engineering

1.1 Importance of Experimental Testing

Natural hazards are the largest potential source of casualties in inhabited areas. Damage to structures causes not only loss of human lives and disruption of lifelines, but also long-term impact on the local, regional and sometimes national and international economies.

One of the main goals of structural and earthquake engineering is to improve the resilience and performance of structures to protect the lives and safety of occupants and control economic losses under an extremely wide range of operational conditions and hazards. Accordingly, the priorities lie in gaining an understanding of the behavior of various classes of structures under different dynamic load types from the elastic range through to developing collapse mechanisms and failure. However, this poses a major challenge as it requires the prediction, with sufficient confidence, of a structure's response beyond design level, all the way to the state of complete collapse.

Today, dynamic analysis of complex structures can be efficiently computed utilizing readily available software. The cost of computation has been continuously reduced, and now very complex and detailed numerical simulations are possible on personal computers. However, for many components or materials, nonlinear behavior and failure modes are still not well understood. In such cases, numerical

---

The original version of this chapter was revised: See the "Chapter Note" section at the end of this chapter for details. The erratum to this chapter is available at https://doi.org/10.1007/978-981-10-5867-7_6

© The Author(s) 2018
R. Al-Mahaidi et al., *Multi-axis Substructure Testing System for Hybrid Simulation*,
SpringerBriefs in Structural Mechanics, https://doi.org/10.1007/978-981-10-5867-7_1

analyses and simulations may not be reliable, since more detailed and complex properties are needed for the critical components to obtain meaningful results. Therefore, laboratory testing remains a necessary tool to improve and validate numerical models over the full range of a structure's response. With this objective in mind, experimental simulations of structures have been conducted to investigate the capacity and failure behavior of various structural systems and critical components that are difficult to model numerically. Based on these studies, the behavior of different structural systems such as multi-story buildings, bridges, coastal structures and others during extreme events is assessed, to enable the design and construction of safer and more resilient structural systems to mitigate natural hazards.

1.2   Experimental Testing Methods

In order to experimentally evaluate the dynamic response of a structure, several techniques can be used in conducting laboratory tests (Filiatrault et al. 2013). The most common method is quasi-static tests that are usually conducted in order to test the behavior of structural components or full-scale structural systems. In this method, the structure is subjected to pre-defined displacement or force history using hydraulic actuators. Typically, these tests are conducted to investigate the hysteretic behavior and capacity of structural components under a cyclic load. Although these tests are fairly easy and economical, they are limited by the predetermined loading protocol. However, in performance-based design, the focus of all decisions is on the demand requirements, the actual behavior of the structural elements and the level of damage during different intensities of extreme loads. Therefore, the predetermined load protocol is generally inadequate for representing the structural behavior, as the load distribution continuously changes during an actual event.

The most realistic approach is the dynamic testing of the entire structure. For example, the use of earthquake shake tables in seismic research provides the means to excite structures in such a way that they reproduce conditions representative of true earthquake ground motions. However, due to the extremely high cost, complexity and damage to the equipment, experimental testing of even a full-scale single-story structure poses significant challenges. The largest shake table in the world, the Hyogo Earthquake Engineering Research Center of Japan (E-Defense) shake table, is located north of Kobe in Miki City, Japan, with dimensions of 15 m by 20 m and the capacity to support building experiments weighing up to 1200 tonnes, which is sufficient to test a full-scale 6-story building. However, not only are these experiments extremely costly, both in terms of operation of a large-scale shake table and in terms of constructing the entire structure on the shake table, they do not provide the large-scale testing environment for tall buildings or horizontally extended structures such as bridges. The George E. Brown Jr. Network for Earthquake Engineering Simulation (NEES) equipment sites in the USA also provide some shake table facilities, such as the twin shake tables in the Buffalo NEES facility, multi-shake table testing in the Nevada NEES facility and a high-performance outdoor shake table in the UC San Diego NEES facility. However, due to limitations on the size and capacity of shake tables, structures are typically tested on a reduced scale or a highly simplified model is used. In particular, in collapse simulation of structures, the shake table test method is expensive, complicated and dangerous, due to the risk associated with the collapse of a structure on the shake table. In addition, a scaled and simplified model does not necessarily represent the response of a full- or large-scale prototype experiencing severe nonlinear deformation and collapse. Scaled specimens can provide a fair understanding of global behavior, but local behavior may not be simulated accurately. However, this local behavior may play a critical role in determining the performance of a structure, given that initial damage usually occurs on a local level. Certain types of behavior, especially local effects such as bond and shear in reinforced concrete members, crack propagation, welding effects and local buckling in steel structures, are well known to have size effects, which casts doubt on the validity of the shake table tests.

The third method is hybrid simulation, also known as pseudo-dynamic testing (Nakashima et al. 1992). Hybrid simulation is a hybrid procedure that combines classical experimental techniques with online computer simulation for cost-effective, large-scale testing of structures under simulated dynamic loads. This method is often called hybrid (rather than pseudo-dynamic testing) since it combines modeling and experiments and can include real dynamic effects in the experiment. According to a report by the US earthquake engineering community, hybrid simulation capabilities are a major emphasis of the next generation of earthquake engineering research (Dyke et al. 2010). However, this role can be expanded to other loading conditions, such as hydrodynamic loading conditions created by waves, traffic and impact loads due to moving vehicles, aerodynamic loads generated by wind, and blast loads. This wide variety of loading conditions can be simulated by incorporating them into the analytical portion of the hybrid model without changing the physical portions of the experiment.

1.3 Hybrid Simulation Fundamentals

Hybrid simulation provides the best advantages of both computational simulation and experimental techniques: the realism of actual testing for the critical components, together with the flexibility and cost-effectiveness of computer modeling. This method is based on domain decomposition, in which the structure of interest can be divided into multiple parts/substructures. On the one hand are the parts and regions that can be reliably modeled in one or more computers, either because of their simple behavior or because they are not considered critical for the analysis conducted. On the other hand are the parts and regions of most interest that are physically tested in one or more laboratories, either because of their highly nonlinear behavior or because they are critical to the safety and performance of the structure. The parts that are numerically simulated are called the numerical or analytical substructures. The parts that are physically modeled and subjected to loads in the laboratory are called the experimental or physical substructures. The combination and interactions of the all substructures form a hybrid model of the complete structure of interest.

Fig. 1.1 Hybrid simulation
technique

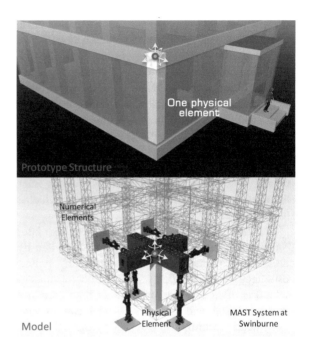

In hybrid testing, the dynamic aspects of the simulation are handled numerically. Therefore, such tests can be viewed as an advanced form of quasi-static testing, where the loading history is determined as the simulation progresses for the structure subjected to a specific dynamic load. The governing equation of the motion is solved similar to pure numerical simulations using a time-stepping integration. The displacement/force demands are then applied to the physical specimen(s), and the resisting forces are measured and fed back to the computation solver to calculate the displacement/force demands corresponding to the next time step.

To illustrate this process for the various types of substructures in hybrid simulation, an example is presented for a multi-story building. Utilizing the hybrid simulation technique, the first-story corner-column, considered the critical element, can be constructed and physically tested in the laboratory, and the remaining parts of the structure, the inertia and damping forces and gravity, dynamic loads and the second-order effects can be reliably modeled in the computer (see Fig. 1.1).

1.4 Advantages and Challenges of Hybrid Simulation

Hybrid simulation provides several advantages, including the following:

1. Hybrid simulation mitigates the errors related to the simplification of the theoretical modeling of complex nonlinear structures or subassemblies by testing

them physically in the laboratory. It is not a trivial task to accurately model complex nonlinear structures, as it requires a series of assumptions that have to be made to simplify the modeling procedure, which is performed at the cost of simulation accuracy and reliability.

2. Hybrid simulation reduces construction/fabrication costs and the overall time for testing in the laboratory. Dynamic testing of the entire structure requires the construction/fabrication of a whole structure, which is an expensive and time-consuming process for a physical test. Since the damage essentially starts as a local phenomenon, hybrid simulation allows the physical testing of only the critical portion of the structure, where the damage is expected.

3. Hybrid simulation reduces uncertainties associated with limited scale of shake table tests by facilitating economical large-scale testing. The size and weight of the physical subassemblies are restricted only by the available laboratory space and the strength of the strong reaction wall/floor. The strength of the specimen is also limited only by the actuator capacities that are available in the laboratory.

4. Hybrid simulation can be conducted on an extended time scale, typically ranging from 100 to 1000 times slower than actual earthquake duration (Carrion and Spencer 2008). This time modification allows the researcher to carefully observe and track the progress of damage throughout the simulation and thus provides important insights into structural component behavior, especially near collapse. Structural performance, such as the failure pattern and initiation of cracks in a special structural region such as beam–column connections, can be closely investigated.

5. Hybrid simulation can be conducted locally or geographically distributed, meaning that individual substructures do not need to be within the same facility, but can be linked by either the Internet or other methods of data transfer. Therefore, laboratories with much larger capacities can be used for experimental subassemblies.

Although hybrid simulation has attracted many researchers in the evaluation of the seismic behavior of structures and offers many advantages, it provides new challenges, as follows:

1. Although hybrid testing is an alternative to shake table testing, the accuracy of the hybrid test is often questioned. To guarantee that the results obtained from a hybrid simulation are valid and reliable, it is important to minimize the contamination of the results by errors. The errors that occur at different stages of a hybrid simulation are modeling errors due to the discretization process, analysis assumptions, numerical errors introduced by the integration and equilibrium solution algorithms, experimental errors generated by the control and transfer systems, and the noise in instrumentation devices and the data acquisition system.

2. Actions on structures during extreme events such as earthquakes are generally multi-directional and continuously varying, due to the time-dependent nature of the input motion. For instance, variations of the axial loads during a seismic excitation may influence the response of the vertical structural components (e.g.,

bridge piers and building columns) since the response of such elements when combined with flexural, shear and torsional actions may differ from the cases when they are not subjected to the same axial load changes. Simulation of such highly coupled multi-directional loading conditions using conventional structural testing methods can be expensive, time-consuming and difficult to achieve. As a result, advanced and innovative experimental techniques and control strategies are under development by researchers (Nakata 2007; Wang et al. 2012; Hashemi et al. 2014; Hashemi and Mosqueda 2014).

3. The experiments should be conducted at a large or full scale to accurately capture the local behavior of the elements. However, the conduct of large-scale experiments may not be feasible, due to the limited resources available in many laboratories, including the number and capability of the actuators available, the dimensions and load capacity of the reaction systems, difficulties in the actuator assemblies and testing configuration in reliably simulating the boundary conditions. Consequently, the specimen may be tested at a small scale or under uni/biaxial loading configurations, which do not necessarily represent the actual action or demand on the structural elements and the corresponding nonlinear response of the prototype system.

4. Conducting multi-directional loading including gravity load effects requires a mixed-mode control strategy. The application of gravity loads has been mainly considered by researchers using a combination of force-control actuators in the vertical direction that are decoupled from displacement-control actuators in the lateral direction of the specimen (Lynn et al. 1996; Pan et al. 2005; Del Carpio Ramos et al. 2015). In these tests, independent of lateral actuators, only the vertical force-control actuators apply the gravity forces, while under large deformations, lateral actuators have a force component in the vertical direction that needs to be accounted for. Therefore, versatile and generally applicable mixed-mode control algorithms are required to take into account instantaneous and spatial coupling in the control systems.

1.5 Objectives and Outline

This manuscript presents the design details and unique capabilities of the MAST system for the hybrid simulation of large-scale structures subjected to extreme dynamic forces. The testing capabilities advance the current state of technology by allowing accurate simulation of complex time-varying 6-DOF boundary effects on large-scale structural components in mixed load/deformation control modes. Utilizing the MAST system, the developments of new materials and structural systems and the effectiveness of new repair/retrofitting strategies can be reliably evaluated using three-dimensional large-scale quasi-static cyclic or local/geographically distributed hybrid simulation tests. The manuscript is organized as follows:

Chapter 2 presents the technical background and literature review on the development of hybrid simulation and summarizes the work by researchers in the fields of substructuring techniques, integration schemes, continuous and real-time hybrid testing, local and geographically distributed hybrid testing and experimental and numerical errors in hybrid testing.

Chapter 3 describes different components of the state-of-the-art system for hybrid simulation at Swinburne, including the design details of the MAST facility, the reaction systems including the strong wall/floor and the cruciform crosshead, servo-hydraulic actuators and the 6-DOF controller system and hybrid simulation architecture.

Chapter 4 presents the results of a range of experiments, including switched/ mixed load/deformation mode quasi-static cyclic and hybrid simulation tests to highlight the unique and powerful capabilities of the MAST system, specifically for the assessment and mitigation of the collapse risk of structures.

Chapter 5 presents a summary of key contributions and concluding remarks. Research areas for further development and study are also briefly discussed.

References

Carrion, J. E., & Spencer, B. F. (2008). Real-time hybrid testing using model-based delay compensation. *Smart Structures and Systems, 4*(6), 809–828.

Del Carpio Ramos, M., Mosqueda, G., & Hashemi, M. J. (2015). Large-scale hybrid simulation of a steel moment frame building structure through collapse. *Journal of Structural Engineering, 142*(1), 04015086.

Dyke, S. J., Stojadinovic, B., Arduino, P., Garlock, M., Luco, N., Ramirez, J. A., et al. (2010). *2020 Vision for earthquake engineering research: Report on an openspace technology workshop on the future of earthquake engineering.* St. Louis, U.S.

Filiatrault, A., Tremblay, R., Christopoulos, C., Folz, B., & Pettinga, D. (2013). *Elements of earthquake engineering and structural dynamics* (3rd ed.). Québec, Canada: Presses Internationales Polytechnique.

Hashemi, M. J., & Mosqueda, G. (2014). Innovative substructuring technique for hybrid simulation of multistory buildings through collapse. *Earthquake Engineering & Structural Dynamics, 43*(14), 2059–2074.

Hashemi, M. J., Masroor, A., & Mosqueda, G. (2014). Implementation of online model updating in hybrid simulation. *Earthquake Engineering and Structural Dynamics, 43*(3), 395–412.

Lynn, A. C., Moehle, J. P., Mahin, S. A., & Holmes, W. T. (1996). Seismic evaluation of existing reinforced concrete building columns. *Earthquake Spectra, 12*(4), 715–739.

Nakashima, M., Kato, H., & Takaoka, E. (1992). Development of real-time pseudo dynamic testing. *Earthquake Engineering and Structural Dynamics, 21*(1), 79–92.

Nakata, N. (2007). Multi-dimensional mixed-mode hybrid simulation, control and applications. Ph.D. Dissertation. U.S.: University of Illinois at Urbana-Champaign.

Pan, P., Nakashima, M., & Tomofuji, H. (2005). Online test using displacement-force mixed control. *Earthquake Engineering and Structural Dynamics, 34*(8), 869–888.

Wang, T., Mosqueda, G., Jacobsen, A., & Cortes-Delgado, M. (2012). Performance evaluation of a distributed hybrid test framework to reproduce the collapse behavior of a structure. *Earthquake Engineering and Structural Dynamics, 41*(2), 295–313.

Chapter 2
Background

Abstract This chapter presents the technical background and literature review on the development of hybrid simulation in the fields of substructuring techniques, integration schemes, continuous and real-time hybrid testing, local and geographically distributed hybrid testing, and experimental and numerical errors in hybrid testing.

Keywords Hybrid simulation · Substructuring · Integration schemes · Continuous testing · Geographically distributed testing · Experimental and numerical errors

2.1 Introduction

The original idea of obtaining the seismic response of a system through a hybrid numerical and experimental model dates back to the late 1960s, when it was first proposed in a Japanese paper by Hakuno et al. (1969). A single-degree-of-freedom (SDOF) cantilever beam was analyzed under seismic loadings using an analog computer in order to solve the equation of motion in combination with an electromagnetic actuator to impose the load on the structure. In order to improve the accuracy of the simulation, the authors suggested using digital computers.

The first major step in the use of digital computers and discrete systems was first taken in the mid-1970s. Takanashi et al. (1975) established the hybrid simulation method in its present form by studying the structural system as a discrete spring-mass system within the time domain. This allowed hybrid simulation to work with typical quasi-static loading systems and provided the necessary foundation to apply hybrid simulation to structural engineering.

Advancements in the development of faster and more reliable testing and computational hardware paved the way for the researchers to expand the capabilities and validation of the hybrid simulation test method. During the late 1970s, 1980s and early 1990s, efforts in Japan and the USA were undertaken in this regard

---

The original version of this chapter was revised: See the "Chapter Note" section at the end of this chapter for details. The erratum to this chapter is available at https://doi.org/10.1007/978-981-10-5867-7_6

that are outlined in Mahin and Shing (1985), Takanashi and Nakashima (1987), Mahin et al. (1989) and Shing et al. (1996). A comprehensive review of the developments in the fields of efficiency, accuracy and performance of hybrid simulation methods can be found in Saouma and Sivaselvan (2008).

2.2 Substructuring Techniques in Hybrid Simulation

Before the mid-1980s, most applications of hybrid simulation required testing of the complete structural system. Consequently, these tests were expensive and required a large-scale testing facility. However, the damage in a structure due to seismic loads could be located within a few critical regions, and as a result, in many cases, it is not necessary to test the entire structural system.

The concept of substructuring, which is similar to the concept of domain decomposition employment in finite-element analysis, is based on splitting the domain of the structure into experimental and numerical substructures and conducting separate analyses on each part, while ensuring the interface constraints are continuously verified both in terms of compatibility and in terms of equilibrium. By using substructuring techniques typically applied to conventional dynamic analysis, the complete structure can be partitioned into several subassemblies. As a result, the parts of a structure that experience complex behavior, which may be difficult to model numerically, are tested physically, while those parts of the structure that have a consistent behavior and are well defined are analyzed numerically.

Dermitzakis and Mahin (1985) suggested utilizing substructuring techniques in order to divide a structure into experimental and numerical subassemblies and perform substructure hybrid simulations. A major advantage of the substructuring technique is that it reduces the space required in order to perform hybrid simulation. Thereby, it facilitates large-scale testing and increases the ability to consider specific local component behavior. Recent advances in substructuring techniques include the implementation of overlapping techniques (Wang et al. 2012; Hashemi and Mosqueda 2014; Del Carpio Ramos et al. 2015) and model updating techniques in hybrid simulation (Hashemi et al. 2014; Elanwar and Elnashai 2015; Shao et al. 2015).

2.3 Time-Integration Algorithms in Hybrid Simulation

One of the most important components of hybrid testing that has a crucial role in the stability and accuracy of the simulation is the numerical integration algorithm. Although there have been many advancements in the development of numerical time-stepping algorithms for pure numerical simulations, most of these methods are not well suited for hybrid simulation. Therefore, there have been many efforts to develop stable, efficient and accurate algorithms for time-integration schemes, specifically for hybrid simulation.

During hybrid simulation, similar to pure finite-element analysis, the equation of motion is discretized in space utilizing elements that are connected with the nodes. This process is carried out to make the system suitable for numerical evaluation and implementation on digital computers. The spatially discretized differential equation can further be simplified utilizing element assembly from local to the global structural DOFs containing all the element contributions:

$$[M]\{\ddot{u}\} + [C]\{\dot{u}\} + [K]\{u\} = \{P\} \tag{2.1}$$

where M, C and K are, respectively, the mass, damping and stiffness matrices assembled from nodal and element properties, $\ddot{u}, \dot{u}$ and u are, respectively, the vectors of nodal accelerations, velocities and displacements for the global DOFs of the structure and P is the vector of system interface and external forces. Note that, while C and K may change during the analysis, M will be regarded as a constant, assuming mass conservation even during failures and collapse.

The equation of motion, which is a second-order ordinary differential equation (ODE), is next discretized in time. This process is performed to advance transient (time-varying) solutions step by step, assuming idealized properties over small time steps. These properties, depending on the scheme considered, are obtained through a set of equations, which can be written in the form of Eqs. 2.2 or 2.3:

$$u_{i+1} = f(u_i, \dot{u}_i, \ddot{u}_i, u_{i-1}, \dot{u}_{i-1}, \ddot{u}_{i-1}, \ldots) \tag{2.2}$$

$$u_{i+1} = f(u_i, \dot{u}_{i+1}, \ddot{u}_{i+1}, u_{i-1}, \dot{u}_i, \ddot{u}_i, \ldots) \tag{2.3}$$

Given that Eq. 2.1 is satisfied with the solution u_i, the task of numerical integration is to advance the solution by finding a displacement increment Δu in such way that Eq. 2.1 is also in equilibrium for $u_{i+1} = u_i + \Delta u$. The various numerical integration schemes can be classified into two types: explicit or implicit.

An explicit scheme, as illustrated in Eq. 2.2, computes the response of the structure at the end of the current time step $(i+1)$, exclusively based on the state of the structure at the beginning of step (i) or earlier. This is an attractive property for hybrid testing, because the actuators are commanded a target displacement without the knowledge of the specimen properties at the target.

An implicit scheme, as illustrated in Eq. 2.3, requires the knowledge of the structural response at the target displacement and is dependent on one or several values from time step $(i+1)$ in order to compute the response. Therefore, an implicit scheme involves a more complex implementation than an explicit one, often including an iterative process or a predictor-corrector algorithm.

Explicit methods are computationally very efficient, easy to implement and fast in their execution. However, the fact that an implicit scheme relies on a future term makes it more stable, regardless of the chosen time step length. In fact, explicit schemes are typically conditionally stable, while implicit schemes can be unconditionally stable (Shing et al. 1996). The implicit schemes, however, require a tangent stiffness matrix, which can be difficult to obtain from physical elements of the hybrid model. In addition, they are often used in combination with an iteration strategy such as the Newton–Raphson algorithm and lead to non-uniform rapidly decreasing displacement increments, which can introduce spurious loading cycles on the physical parts of the hybrid model. In order to address these issues, integration methods have been introduced to apply the implicit iterations only in numerical substructure, while using the initial-elastic stiffness matrix for the experimental substructure to approximate its behavior (Dermitzakis and Mahin 1985; Nakashima et al. 1990; Schellenberg et al. 2009).

2.4 Continuous Hybrid Simulation

In conventional pseudo-dynamic testing methods, the load is applied to the specimen using a ramp-hold procedure. In continuous hybrid simulation testing, the load is applied smoothly, without starts and stops. There are two main reasons for running the simulation continuously. The first problem associated with ramp-hold loading is that during the hold period, reductions in the restoring force can occur due to force relaxation. Force relaxation is caused by stress relaxation, which is the decay in stress over time while the strain is held constant. Another reason is that the modern servo-hydraulic controller systems run at the sampling rate of 1024 Hz and above. As a result, signal generation for actuator commands should be performed at this sampling rate, which is deterministic. On the other hand, in hybrid testing there are inevitable time lags and delays that are generally non-deterministic. Therefore, these two processes should be synchronized to avoid systematic errors and also enable the achievement of a smooth loading for the test structure and a reduction in overall testing time. For this purpose, a predictor-corrector command generation algorithm is placed between computation solver and the controller to generate the command displacement for the actuator controller, while the computation driver solves the equation of motion. Once the target displacement is computed, the algorithm corrects the command displacement path toward the target displacement.

Takanashi and Ohi (1983) first introduced the concept of continuous loading and fast hybrid testing. Mosqueda et al. (2005) presented a system for continuous hybrid simulation with distributed experimental sites connected through the Internet. Since the time required for network communication is random, a solution using an event-driven controller was proposed. The resulting system was an event-driven version of the system proposed by Nakashima and Masaoka (1999), in which the tasks of integration of the equation of motion and signal generation run as two different processes.

2.5 Real-Time Hybrid Simulation

Hybrid simulation does not require dynamic loading, since dynamic effects such as inertia and damping forces are considered in the numerical portion of the equations of motion. However, the development of velocity-dependent structural components and devices to control the response of structures caused researchers to seek to expand the capabilities of hybrid simulation to work in real time.

Real-time hybrid simulation has been proposed to fully capture strain rate, damping and inertial effects by computing each numerical integration time step of the experiment in exactly that amount of time. Studies on real-time hybrid simulation began in the early 1990s and have continued to the present as more velocity-dependent systems are applied to structures. The major difference between real-time hybrid simulation and quasi-static hybrid simulation is that in addition to displacements, velocities are controlled for the experimental portion of the test.

Nakashima (2001) presented an overview of the development of real-time hybrid simulation systems. To conduct real-time testing, it was essential to develop a procedure that allowed for continuous real-time loading without interruption of the displacement signals sent to the digital controller. While the initial tests (Horiuchi et al. 1999; Nakashima and Masaoka 1999) dealt only with SDOF systems, more difficulties arise as a result of controlling the multiple actuators needed for MDOF tested structures. As a result, many of the hybrid simulation studies focused on resolving some of these limitations (Reinhorn et al. 2004; Shing et al. 2004; Bonnet 2006; Shao and Reinhorn 2012). Real-time hybrid simulations have been successfully conducted for the investigation of the dynamic behavior of structures with rate-dependent devices (Wu et al. 2007; Carrion and Spencer 2008; Karavasilis et al. 2011; Chen and Ricles 2012).

2.6   Geographically Distributed Hybrid Simulation

The popularity of hybrid simulation among structural engineering researchers has grown to a great extent. Geographically distributed testing is one recent concept that has been developed from the use of substructuring techniques and benefited from technological advances in data transfer and computing.

The concept of geographically distributed testing is that individual substructures do not need to be within the same facility, but can be linked by either the Internet or another methods of data transfer. By breaking a model into selected subassemblies and distributing them within a network of laboratories and computational sites, a researcher is able to take advantage of different capabilities available at the various facilities.

Campbell and Stojadinovic (1998) first suggested the geographical distribution of structural subassemblies within a network of laboratories, where the individual sites are connected through the Internet. Mosqueda et al. (2005) developed a three-loop architecture that allowed for the first time the execution of continuous, geographically distributed hybrid simulations with multiple subassemblies. Compared to previous geographically distributed hybrid simulations that utilized a hold-and-ramp loading procedure, it was possible to significantly reduce execution times and eliminate force relaxation problems. Kim et al. (2012) presented a framework used to successfully conduct geographically distributed real-time hybrid simulation tests.

2.7 Error Propagation in Hybrid Simulation

While hybrid simulation is an attractive test method, it is prone to both numerical and experimental errors that must be carefully addressed to achieve reliable results (Ahmadizadeh and Mosqueda 2009). For example, in substructure hybrid simulations, a significant portion of a structure is typically modeled numerically, with the simulation result being highly dependent on the performance and stability of the computation. Experimental errors can also be introduced into the simulation mainly through two sources of error: (1) errors generated by the difference in the imposed displacement versus the computed displacement or (2) errors generated by incorrect force measurements from the experimental substructure, which is then used to solve the equation of motion. Since hybrid simulation is a closed-loop system and also a stepwise process, these errors can accumulate, resulting in an overall decrease in the accuracy of the hybrid simulation and sometimes instability of the simulation (Hashemi et al. 2016a, b). Implementation of simplified or approximate substructuring techniques may also contribute in the form of modeling errors. Small errors can accumulate during the experiment and significantly affect the simulation results.

The propagation of random and systematic errors in hybrid simulation has been thoroughly studied. Shing and Mahin (1983), Nakashima et al. (1985), and Thewalt and Mahin (1987) provided significant contributions to identifying and determining the characteristics of experimental errors within hybrid simulation tests. It was found that systematic overshoot error increases the apparent damping of the system, while systematic undershoot results in negative damping and can produce an increase in the response, particularly corresponding to the higher-frequency response modes. Mosqueda et al. (2005), Ahmadizadeh and Mosqueda (2009) and Hashemi et al. (2016a, b), provide detailed explanations and summaries on errors in hybrid simulations, including errors based on modeling, implementation techniques and the experimental setup.

2.8 Collapse Simulation Through Hybrid Testing

Understanding and modeling structural failure under dynamic loadings remain a difficult challenge in structural engineering. Specifically, structural behavior through collapse has become increasingly important for applications in performance-based design. Although experimental simulations of structures have been conducted to investigate the seismic capacity of various structural systems and critical components, few hybrid tests have examined structures up to collapse with significant geometric and material nonlinearities (Schellenberg et al. 2008a, b; Shoraka et al. 2008; Wang et al. 2008, 2012; Del Carpio et al. 2014; Hashemi and Mosqueda 2014; Hashemi et al. 2016a, b, 2017).

Collapse simulation of large-/full-scale structures is not always feasible due to space requirements and high costs. On the other hand, reduced-scale experiments are not always reliable due to the difficulties in reproducing local behavior such as connection details. Wang et al. (2008) simulated the seismic behavior of a one-bay, four-story steel moment frame through collapse. The simulation was geographically distributed with substructuring, considering the column bases as the experimental portion, while the superstructure was analyzed numerically.

In an effort to show the potential of hybrid testing to simulate structural behavior through collapse, Wang et al. (2012) conducted a geographically distributed hybrid test that reproduced the collapse behavior of a four-story, two-bay, steel moment frame, previously tested at the Hyogo Earthquake Engineering Research Center of Japan (E-Defence). The hybrid testing was capable of successfully tracing the response of the structure to collapse.

Hashemi and Mosqueda (2014) proposed a framework for collapse simulation of complex structures through hybrid testing that facilitates the system-level experimental testing of the structures with distributed damage through collapse. This framework uses an advanced substructuring technique that handles the interface between a complex numerical model and the physical subassembly through the use of additional sensing in the feedback loop to obtain internal member forces. This framework was later used in another study to investigate the seismic response of large-scale steel gravity and moment frames through collapse (Del Carpio Ramos et al. 2015).

References

Ahmadizadeh, M., & Mosqueda, G. (2009). Online energy-based error indicator for the assessment of numerical and experimental errors in a hybrid simulation. *Engineering Structures, 31*(9), 1987–1996.

Bonnet, P. A. (2006). *The development of multi-axis real-time substructure testing.* Ph.D. Dissertation. London, UK: University of Oxford.

Campbell, S., & Stojadinovic, B. (1998). A system for simultaneous pseudodynamic testing of multiple substructures. In *Sixth U.S. National Conference on Earthquake Engineering.* Seattle, U.S.

Carrion, J. E., & Spencer, B. F. (2008). Real-time hybrid testing using model-based delay compensation. *Smart Structures and Systems, 4*(6), 809–828.

Chen, C., & Ricles, J. M. (2012). Large-scale real-time hybrid simulation involving multiple experimental substructures and adaptive actuator delay compensation. *Earthquake Engineering and Structural Dynamics, 41*(3), 549–569.

Del Carpio, M., Mosqueda, G., & Lignos, D. G. (2014). *Hybrid simulation of the seismic response of a steel moment frame building structure through collapse.* Multidisciplinary Center for Earthquake Engineering Research, University at Buffalo, U.S.

Del Carpio Ramos, M., Mosqueda, G., & Hashemi, M. J. (2015). Large-scale hybrid simulation of a steel moment frame building structure through collapse. *Journal of Structural Engineering, 142*(1), 04015086.

Dermitzakis, S. N., & Mahin, S. A. (1985) *Development of substructuring techniques for on-line computer controlled seismic performance testing.* Berkeley, U.S.: Earthquake Engineering Research Center, University of California.

Elanwar, H. H., & Elnashai, A. S. (2015). Framework for online model updating in earthquake hybrid simulations. *Journal of Earthquake Engineering.* doi:10.1080/13632469.2015.1051637.

Hakuno, M., Shidawara, M., & Hara, T. (1969). Dynamic destructive test of a cantilever beam, controlled by an analog-computer. *Proceedings of the Japan Society of Civil Engineers, 171,* 1–9.

Hashemi, M. J., Al-Ogaidi, Y., Al-Mahaidi, R., Kalfat, R., Tsang, H., & Wilson, J. (2016a). Application of hybrid simulation for collapse assessment of post-earthquake CFRP-repaired RC columns. *Journal of Structural Engineering, 143*(1).

Hashemi, M. J., Masroor, A., & Mosqueda, G. (2014). Implementation of online model updating in hybrid simulation. *Earthquake Engineering and Structural Dynamics, 43*(3), 395–412.

Hashemi, M. J., & Mosqueda, G. (2014). Innovative substructuring technique for hybrid simulation of multistory buildings through collapse. *Earthquake Engineering and Structural Dynamics, 43*(14), 2059–2074.

Hashemi, M. J., Mosqueda, G., Lignos, D. G., Medina, R. A., & Miranda, E. (2016b). Assessment of numerical and experimental errors in hybrid simulation of framed structural systems through collapse. *Journal of Earthquake Engineering, 20*(6), 885–909.

Hashemi, M. J., Tsang, H. H., Al-Ogaidi, Y., Wilson, J. L., & Al-Mahaidi, R. (2017). Collapse Assessment of Reinforced Concrete Building Columns through Multi-Axis Hybrid Simulation, *Structural Journal, 114*(02).

Horiuchi, T., Inoue, M., Konno, T., & Namita, Y. (1999). Real-time hybrid experimental system with actuator delay compensation and its application to a piping system with energy absorber. *Earthquake Engineering and Structural Dynamics, 28*(10), 1121–1141.

Karavasilis, T. L., Ricles, J. M., Sause, R., & Chen, C. (2011). Experimental evaluation of the seismic performance of steel MRFs with compressed elastomer dampers using large-scale real-time hybrid simulation. *Engineering Structures, 33*(6), 1859–1869.

Kim, S. J., Christenson, R., Phillips, B., & Spencer, B. F. (2012). Geographically distributed real-time hybrid simulation of mr dampers for seismic hazard mitigation. In *20th Analysis & Computation Specialty Conference*. Chicago, U.S.

Mahin, S. A., & Shing, P. S. B. (1985). Pseudodynamic Method for Seismic Testing. *Journal of Structural Engineering, 111*(7), 1482–1503.

Mahin, S. A., Shing, P. S. B., Thewalt, C. R., & Hanson, R. D. (1989). Pseudodynamic Test Method-Current Status and Future-Directions. *Journal of Structural Engineering-Asce, 115*(8), 2113–2128.

Mosqueda, G., Stojadinovic, B., & Mahin, S. (2005). *Implementation and accuracy of continuous hybrid simulation with geographically distributed substructures*. U.S.: Earthquake Engineering Research Center, University of California Berkeley.

Nakashima, M. (2001). Development, potential, and limitations of real-time online (pseudo-dynamic) testing. *Philosophical Transactions of the Royal Society of London Series A-Mathematical Physical and Engineering Sciences, 359*(1786), 1851–1867.

Nakashima, M., Kaminosono, T., & Ishida, M. (1990). Integration techniques for substructure pseudodynamic test. In *4th U.S. National Conference on Earthquake Engineering*. California: Palm Springs.

Nakashima, M., Kato, H., & Kaminosono, T. (1985). Simulation of earthquake response by pseudo dynamic (PSD) testing technique (part 3 estimation of response errors caused by PSD test control errors. Annual Meeting of the Architectural Institute of Japan, Tokyo, Japan.

Nakashima, M., & Masaoka, N. (1999). Real-time on-line test for MDOF systems. *Earthquake Engineering and Structural Dynamics, 28*(4), 393–420.

Reinhorn, A. M., Sivaselvan, M. V., Liang, X., & Shao, X. Y. (2004). Real-time dynamic hybrid testing of structural systems. In *13th World Conference on Earthquake Engineering*. Vancouver, Canada.

Saouma, V., & Sivaselvan, M. (2008). *Hybrid simulation: Theory, implementation and applications*. London, UK: Taylor & Francis Group.

Schellenberg, A., Huang, Y., & Mahin, S. A. (2008a). Structural FE-Software Coupling through the Experimental Software Framework, OpenFresco. In *14th World Conference on Earthquake Engineering*, Beijing, China.

Schellenberg, A. H., Mahin, S. A., Yang, T., Mahin, S., & Stojadinovic, B. (2008b). Hybrid simulation of structural collapse. In *14th World Conference on Earthquake Engineering*, Beijing, China.

Schellenberg, A. H., Mahin, S. A., & Fenves, G. L. (2009). *Advanced implementation of hybrid simulation*. Berkeley, U.S.: Pacific Earthquake Engineering Research Center, University of California.

Shao, X., Mueller, A., & Mohammed, B. (2015). Real-time hybrid simulation with online model updating: Methodology and implementation. *Journal of Engineering Mechanics (ASCE)*. doi:10.1061/(ASCE)EM.1943-7889.0000987.

Shao, X., & Reinhorn, A. M. (2012). Development of a controller platform for force-based real-time hybrid simulation. *Journal of Earthquake Engineering, 16*(2), 274–295.

Shing, P. B., & Mahin, S. A. (1983). *Experimental error propagation in pseudodynamic testing*. Earthquake Engineering Research Center, University of California Berkeley, U.S.

Shing, P. B., Nakashima, M., & Bursi, O. S. (1996). Application of pseudodynamic test method to structural research. *Earthquake Spectra, 12*(1), 29–56.

Shing, P. B., Wei, Z., Jung, R.-Y., & Stauffer, E. (2004). NEES fast hybrid test system at the University of Colorado. In *13th World Conference on Earthquake Engineering*. Vancouver, Canada.

Shoraka, M. B., Charlet, A. Y., Elwood, K. J., & Haukaas, T. (2008). Hybrid simulation of the gravity load collapse of reinforced concrete frames. In *18th Analysis and Computation Specialty Conference*, Vancouver, Canada.

Takanashi, K., & Ohi, K. (1983). *Earthquake response analysis of steel structures by rapid computer-actuator on-line system, (1) a progress report, trial system and dynamic response of steel beams*. Tokyo, Japan: Bull. Earthquake Resistant Struct. Research Center (ERS), Inst. of Industrial Sci., Univ. of Tokyo.

Takanashi, K., & Nakashima, M. (1987). Japanese Activities on On-Line Testing. *Journal of Engineering Mechanics, 113*(7), 1014–1032.

Takanashi, K., Udagawa, K., Seki, M., Okada, T., & Tanaka, H. (1975) Nonlinear earthquake response analysis of structures by a computer-actuator on-line system, *Bulletin of Earthquake Resistant Structure Research Centre* (No. 8). Japan: Institute of Industrial Science, University of Tokyo.

Thewalt, C. R., & Mahin, S. A. (1987). *Hybrid solution techniques for generalized pseudodynamic testing*. Earthquake Engineering Research Center, University of California Berkeley, U.S.

Wang, T., McCormick, J., Yoshitake, N., Pan, P., Murata, Y., & Nakashima, M. (2008). Collapse simulation of a four-story steel moment frame by a distributed online hybrid test. *Earthquake Engineering & Structural Dynamics, 37*(6), 955–974.

Wang, T., Mosqueda, G., Jacobsen, A., & Cortes-Delgado, M. (2012). Performance evaluation of a distributed hybrid test framework to reproduce the collapse behavior of a structure. *Earthquake Engineering & Structural Dynamics, 41*(2), 295–313.

Wu, B., Wang, Q. Y., Shing, P. B., & Ou, J. P. (2007). Equivalent force control method for generalized real-time substructure testing with implicit integration. *Earthquake Engineering and Structural Dynamics, 36*(9), 1127–1149.

Chapter 3
State-of-the-Art System for Hybrid Simulation at Swinburne

Abstract This chapter describes the different components of the state-of-the-art system for hybrid simulation at Swinburne, including the design details of the MAST facility, the reaction systems including the strong wall/floor and the cruciform crosshead, servo-hydraulic actuators and the 6-DOF controller system, and hybrid simulation architecture.

Keywords MAST system · 6-DOF hybrid testing · Servo-hydraulic actuators · Control system

3.1 Introduction

The 6-DOF loading on structural components has been performed previously in the George E. Brown Jr. Network for Earthquake Engineering Simulation (NEES) facilities in the USA, including the Multi-Axial Subassemblage Testing Laboratory located at the University of Minnesota, Minneapolis (French et al. 2004), which has been used in quasi-static tests, and the Multi-Axial Full-Scale Sub-Structure Testing and Simulation facility at the University of Illinois at Urbana-Champaign (Kim et al. 2011; Mahmoud et al. 2013), which has been used in displacement-control hybrid simulation experiments (see Fig. 3.1). These systems have the capacity for large-scale testing and the ability to control multiple DOFs at the boundaries of physical specimens. Building on the same concept, the Multi-Axis Sub-structure Testing (MAST) system at Swinburne (see Fig. 3.2) has been developed to advance the current state of technology by allowing accurate simulation of complex time-varying 6-DOF boundary effects on large-scale structural components in mixed load/deformation control modes during hybrid simulation (Hashemi et al. 2015).

The unique and versatile capabilities of the MAST system will greatly expand the experimental testing of large-scale structural components such as beam–column

---

The original version of this chapter was revised: See the "Chapter Note" section at the end of this chapter for details. The erratum to this chapter is available at https://doi.org/10.1007/978-981-10-5867-7_6

© The Author(s) 2018
R. Al-Mahaidi et al., *Multi-axis Substructure Testing System for Hybrid Simulation*,
SpringerBriefs in Structural Mechanics, https://doi.org/10.1007/978-981-10-5867-7_3

(a)   (b)

Fig. 3.1 Multi-directional loading at NEES facilities: **a** Multi-Axial Subassemblage Testing Laboratory at Minnesota, USA. **b** Multi-Axial Full-Scale Sub-Structure Testing and Simulation Laboratory at Illinois, USA

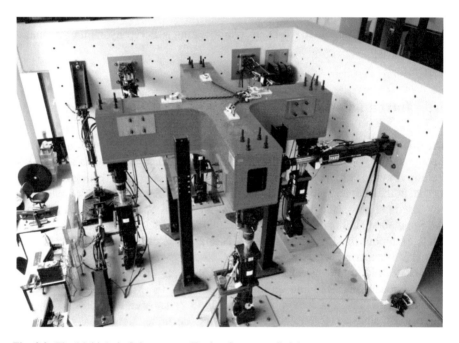

Fig. 3.2 The Multi-Axis Sub-structure Testing System at Swinburne

frame systems, walls and bridge piers. Using the MAST system, the development of new materials and structural components and the effectiveness of new repair/ retrofitting strategies for seismically damaged structural elements can be reliably evaluated through three-dimensional large-scale local/geographically distributed hybrid simulation, which provides significant insight into the effects of extreme loading events on civil structures.

The key components of the MAST facility are:

1. Four ± 1MN vertical hydraulic actuators as well as two pairs of ± 500 kN horizontal actuators in orthogonal directions. Auxiliary actuators are also available for additional loading configurations on the specimen.
2. A 9.5-tonne steel crosshead that transfers the 6-DOF forces from the actuators to the specimen. The test area under the crosshead is approximately 3 m × 3 m in-plan and 3.2 m high.
3. A reaction system comprising an L-shaped strong wall (5 m tall × 1 m thick) and a 1-m-thick strong floor.
4. An advanced servo-hydraulic control system capable of imposing simultaneous 6-DOF states of deformation and load in switched and mixed-mode control. Also, the center of rotation (COR) (i.e., the fixed point around which the 6-DOF movements of the control point occurs) can be relocated and/or reoriented by assigning the desired values.
5. An advanced three-loop hybrid simulation architecture (Stojadinovic et al. 2006) including an innermost servo-valve control loop that runs on a MTS Flex Test 100 Controller (MTS Systems Corporation 2014), a middle actuator command generation loop that runs on the xPC-Target real-time digital signal processor (DSP) and includes the predictor-corrector algorithm and an outer integrator loop that runs on the xPC-Host and includes OpenSees (McKenna 2011), OpenFresco (Schellenberg et al. 2009) and MATLAB/Simulink (The MathWorks Inc. 2014).
6. Additional high-precision draw-wire absolute encoders (SICK 2014) with a resolution of 25 μ that can be directly fed back to the controller.

Fig. 3.3 Smart Structures Laboratory at Swinburne University of Technology

Fig. 3.4 Advanced Technology Center at Swinburne University of Technology

The MAST system is located in the Smart Structures Laboratory (see Fig. 3.3) at Swinburne's architecturally striking Advanced Technologies Center (ATC) (see Fig. 3.4). The $15 million laboratory is a major 3-D testing facility developed for large-scale testing of civil, mechanical, aerospace and mining engineering components and systems, and the only one of its type is available in Australia. The 1.0-m-thick strong floor measures 20 m × 8 m in-plan with two 5-m-tall reaction walls meeting at one corner. The 3-D strong cell contains a grid of tie-down points 0.5 m apart to secure the test specimens in place, in addition to a suite of hydraulic actuators and universal testing machines varying in capacity from 10 to 500 tonnes. The laboratory is serviced by adjacent workshops and a hydraulic pump system located in the basement. The facility is housed in a large architecturally designed test hall about 8 m tall with a 10-tonne crane.

3.2 MAST Reaction Systems

3.2.1 Design of the Strong Wall/Floor

A strong floor and a reaction wall were designed to provide supports for vertical and horizontal actuators for the testing of full-scale structures. The primary objective of the facility is to simulate earthquake loading of large-scale structural elements such as bridge piers, bridge girders and components of multi-story buildings when subjected to static and pseudo-dynamic testing. The design of the facility was influenced by the following performance objectives:

1. The dimensions and overall size of the strong floor would limit the types of tests that could be conducted in the facility due to size constraints.

2. The thickness of the strong wall/floor and the quantity of steel reinforcement and post-tensioning would limit the loads that could be applied to the structure and strongly influence the size/type of actuators that could be used.
3. No cracking of the reinforced concrete strong wall/floor was permitted under service loading conditions. In this case, both the wall and floor were chosen to be post-tensioned to increase the serviceability performance of the structure and to ensure that tensile stresses within the concrete are maintained at levels below the characteristic tensile strength of the concrete.
4. The dimensions of the strong wall/floor need to be sufficient in size and provide adequate stiffness to allow the application of large horizontal forces while maintaining negligible deformations in the wall/floor itself.

In accordance with the above performance objectives, the design of the strong floor was undertaken by Waterman International Consulting (Waterman Group plc 2010) engineers in collaboration with Swinburne University of Technology. The overall dimensions of the facility are as follows:

1. The strong floor is 21 m × 8 m in-plan with thickness of 1 m.
2. The L-shaped strong wall is 5 m high, 1 m thick and, respectively, 8 and 6 m long on each side.

The final configuration of the facility showing all the dimensions is shown in Fig. 3.5.

   To evaluate whether the proposed geometry of the facility was acceptable in meeting the proposed performance objectives, a solid 3-D FE model was constructed using brick elements in the program ANSYS Inc. (2012). Concrete is considered to be a quasi-brittle material capable of both cracking and crushing behaviors under tensile and compressive stress. The compressive response of concrete is highly nonlinear, whereas in tension, the stress–strain response is approximately linear elastic up to the maximum tensile strength, resulting in cracking and a sudden loss of strength. However, since one of the performance objectives was to ensure that the tensile strength of the concrete should not be exceeded, the definition of linear elastic properties for the concrete material was

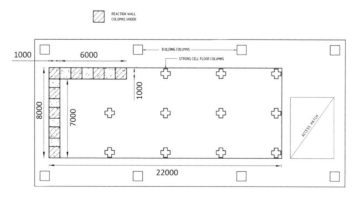

Fig. 3.5 Final configuration of the strong floor/wall system

Table 3.1 Material properties of concrete used in FE model

Property	Value
Density	2400 (kg m^{-3})
Young's modulus	39,117 (MPa)
Poisson's ratio	0.2
Characteristic tensile strength	3.87 (MPa)
Characteristic compressive strength	60 (MPa)

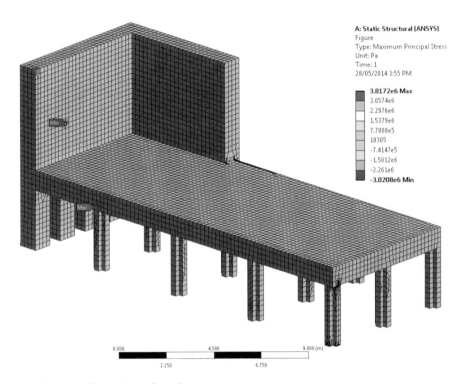

Fig. 3.6 Final FE model configuration

adopted for simplicity. Table 3.1 summarizes the material properties of the concrete used in the FE model.

The model was restrained by fixing translations in x, y and z for the base of the columns. The resultant model and mesh detail are presented in Fig. 3.6

In order to meet the objectives of the performance criteria, over 100 load cases were constructed, representing all possible actuator positions on the reaction wall. In each load case, an iterative procedure was used to increase the actuator loads until the tensile strength of the concrete was reached. The maximum load reached for each load case, and the respective loading key plan is presented in Figs. 3.7, 3.8, 3.9, 3.10 and Table 3.2.

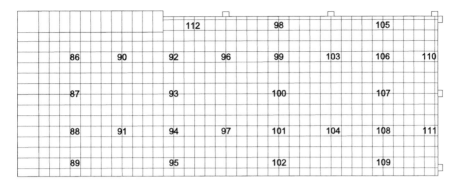

Fig. 3.7 Load-case key (plan view)

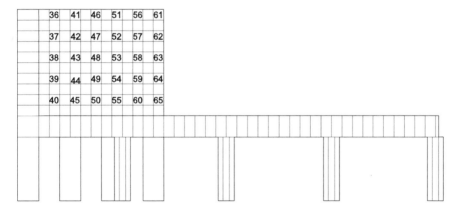

Fig. 3.8 Load-case key (south elevation)

Although the actuators and crosshead have the capability to exert a total vertical force of 1 MN and 500 kN (per actuator) to the strong floor and strong wall, respectively, it is clear from Table 3.2 that the permissible load may control the maximum actuator loads that may be used in any testing configuration. Therefore, any proposed testing should be designed with the permissible actuator loads in mind to ensure that the strong floor and reaction wall are not overloaded beyond acceptable limits. An overview of the strong wall/floor is shown in Fig. 3.11.

3.2.2   Design of the Steel Crosshead

The 6-DOF forces from the actuators are transferred to the testing specimen through a rigid steel crosshead. The actuators move the crosshead to apply the desired loads and deformations. Preliminary designs for the crosshead were developed based on

Fig. 3.9 Load-case key (east elevation)

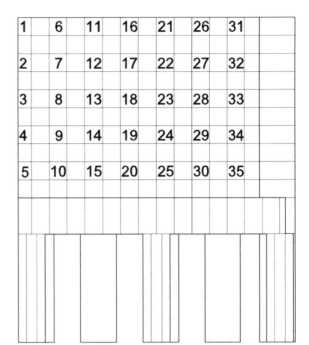

the design criteria, which resulted in two alternative crosshead sizes studied in detail using finite-element (FE) simulations. The following performance objectives were used in the design of the crosshead:

1. The size of the cross section, thickness of steel plates and weld sizes should be sufficient to resist a maximum vertical load of 4 MN (applied via four 1 MN actuators) and horizontal biaxial load of 1 MN (applied via two pairs of 500 kN actuators in orthogonal directions), while ensuring that steel stresses are maintained below acceptable limits.
2. The mounting holes at the bottom of the crosshead must allow maximum flexibility for attaching test structures.
3. The mounting holes and loading plates for the horizontal actuators must be based on 500 kN capacity actuators.
4. The total weight of the crosshead should be less than 10 tonnes, which is the maximum permissible load carrying capacity of the overhead crane in the laboratory.

In accordance with the above performance objectives, the design of the crosshead in the form of a cruciform was undertaken by Hofmann Engineering Pty. Ltd. (Hofmann 2013) in collaboration with Swinburne University of Technology. The first crosshead size investigated consisted of a 650 mm × 800 mm cross section for each crosshead arm and a total length of 4750 mm. However, a preliminary FE study proved that the cross section was insufficient in size to accommodate the loading conditions described above, due to the violation of acceptable stress limits.

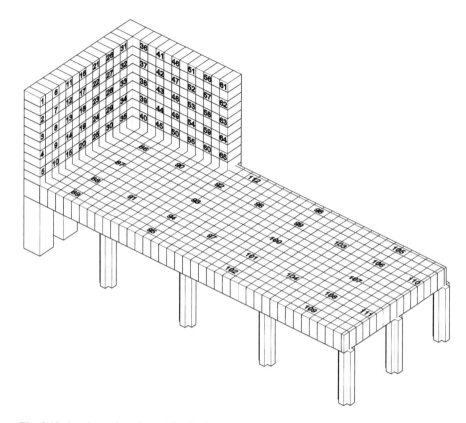

Fig. 3.10 Load-case key (isometric view)

Further, FE simulations were conducted to determine the optimal size of the crosshead arms, resulting in a final cross-sectional dimension of 800 mm × 1010 mm and a total length of 5020 mm. Based on the above analysis, detailed drawings were produced for manufacture and more detailed FE study was conducted to further verify the strength of connections and welds. The final specifications of the crosshead included:

1. 60 mm nominal steel thickness for bottom vertical actuator base plates and 50 mm thickness under side horizontal actuator loading plates.
2. 32 mm bottom plate thickness and 20-mm-thick walls to distribute bearing stress.
3. Double bottom plate thickness in the center of the cruciform to reduce peak torsional stress.
4. Grade 420-MPa steel.
5. Curved stiffener plates to reduce stress concentrations at corners.
6. Holes in internal stiffener plates to reduce weight, resulting in a total weight of 9470 kg.

Table 3.2 Maximum allowable actuator loads for given load case and actuator positions

Load case	Combination	Max allowed load (kN)	Load case	Combination	Max allowed load (kN)
LC1	(1)	350	LC41	(51)	605
LC2	(2)	380	LC42	(52)	860
LC3	(3)	415	LC43	(53)	940
LC4	(4)	450	LC44	(54)	1075
LC5	(5)	490	LC45	(55)	1270
LC6	(6)	440	LC46	(56)	575
LC7	(7)	485	LC47	(57)	650
LC8	(8)	550	LC48	(58)	775
LC9	(9)	675	LC49	(59)	875
LC10	(10)	885	LC50	(60)	1050
LC11	(11)	575	LC51	(61)	350
LC12	(12)	650	LC52	(62)	380
LC13	(13)	775	LC53	(63)	415
LC14	(14)	875	LC54	(64)	450
LC15	(15)	1050	LC55	(65)	490
LC16	(16)	605	LC56	(1) + (61)	(285) + (285)
LC17	(17)	860	LC57	(2) + (62)	(312.5) + (312.5)
LC18	(18)	940	LC58	(3) + (63)	(350) + (350)
LC19	(19)	1075	LC59	(4) + (64)	(415) + (415)
LC20	(20)	1270	LC60	(5) + (65)	(550) + (550)
LC21	(21)	600	LC61	(6) + (61)	(280) + (280)
LC22	(22)	900	LC62	(7) + (62)	(310) + (310)
LC23	(23)	1150	LC63	(8) + (63)	(350) + (350)
LC24	(24)	1325	LC64	(9) + (64)	(415) + (415)
LC25	(25)	1425	LC65	(10) + (65)	(550) + (550)
LC26	(26)	615	LC66	(11) + (56)	(310) + (310)
LC27	(27)	1070	LC67	(12) + (57)	(385) + (385)
LC28	(28)	1325	LC68	(13) + (58)	(450) + (450)
LC29	(29)	1410	LC69	(14) + (59)	(565) + (565)
LC30	(30)	1465	LC70	(15) + (60)	(925) + (925)
LC31	(41)	615	LC71	(16) + (51)	(310) + (310)
LC32	(42)	1070	LC72	(17) + (52)	(435) + (435)
LC33	(43)	1325	LC73	(18) + (53)	(600) + (600)
LC34	(44)	1410	LC74	(19) + (54)	(825) + (825)
LC35	(45)	1465	LC75	(20) + (55)	(1000) + (1000)
LC36	(46)	600	LC76	(21) + (46)	315
LC37	(47)	900	LC77	(22) + (47)	475

(continued)

Table 3.2 (continued)

Load case	Combination	Max allowed load (kN)	Load case	Combination	Max allowed load (kN)
LC38	(48)	1150	LC78	(23) + (48)	800
LC39	(49)	1325	LC79	(24) + (49)	950
LC40	(50)	1425	LC80	(25) + (50)	1075
LC81	(26) + (41)	335	LC99	(99)	2750
LC82	(27) + (42)	575	LC100	(100)	2950
LC83	(28) + (43)	950	LC101	(101)	2700
LC84	(29) + (44)	1145	LC102	(102)	1625
LC85	(30) + (45)	1145	LC103	(103)	2250
LC86	(86)	3100	LC104	(104)	2950
LC87	(87)	3100	LC105	(105)	1381
LC88	(88)	2700	LC106	(106)	2338
LC89	(89)	1650	LC107	(107)	2508
LC90	(90)	3350	LC108	(108)	2330
LC91	(91)	2950	LC109	(109)	1381
LC92	(92)	1950	LC110	(110)	1381
LC93	(93)	3075	LC111	(111)	1640
LC94	(94)	2750	LC112	(110) + (111)	1750
LC95	(95)	1625	LC113	(106) + (108)	1950
LC96	(96)	2250	LC114	(103) + (104)	2950
LC97	(97)	2950	LC115	(101) + (108)	(850) + (850)
LC98	(98)	1625	LC116	(100) + (107)	(400) + (400)

Design specifications and details of the crosshead are presented in Appendix Figures A.1 to A.3. Finite-element modeling for the crosshead was undertaken by building a 3-D model that reflected the geometries and material properties to be used in manufacture. The model included all relevant details, such as holes for base plate connections and stiffener plates. Zones of weakness at weld connections were considered by locally modeling elements of lower strength/stiffness in the vicinity of welds. The analysis was geometrically nonlinear.

Restraint was applied to the base plate of the crosshead by fixing the translation in the z-direction of the base plate. In addition, the perimeters of the holes were fixed in the x, y, z directions to reflect the restraining effects of the bolts. The appropriate load cases considered are outlined in Figs. 3.12 and 3.13. Two load cases were considered to induce the highest possible flexure and torsion within the structure. Note that in all figures that follow, the 32-mm machined bottom steel surface of the crosshead is shown topside for clarity. Compressive load was applied as a uniformly distributed load over the area of the loading plates in contact with the actuators. Tensile forces were applied to an area around each loading plate hole using the appropriate washer size.

Fig. 3.11 Strong wall/floor in the Smart Structure Laboratory at Swinburne

Fig. 3.12 MAST sample load case for maximum torsion

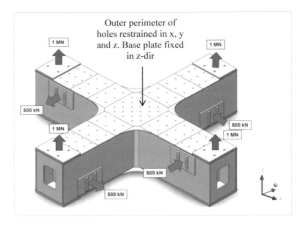

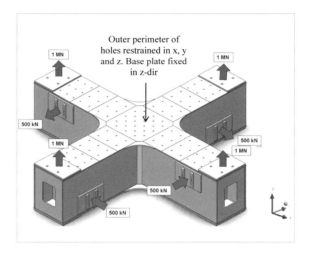

Fig. 3.13 MAST sample load case for maximum flexure

Load case 2 was found to produce the highest stress concentrations around the outer perimeter of bolt holes attached to the bottom base plate. The design initially used a bottom base plate thickness of 40 mm, which was later increased to 60 mm to ensure that the stresses in the vicinity of the holes did not exceed the yield strength of steel. The results are presented in Fig. 3.14, which highlights a maximum stress of 358 MPa, approximately 85% of yield. Measures were taken to

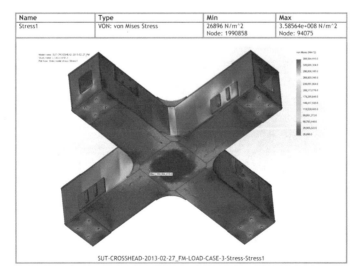

Fig. 3.14 Von Mises stress contour for FEA load case 2

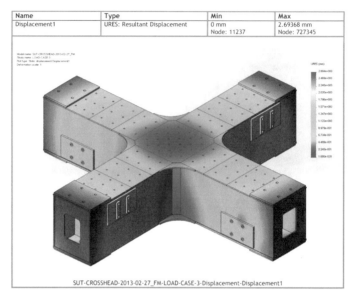

Name	Type	Min	Max
Displacement1	URES: Resultant Displacement	0 mm Node: 11237	2.69368 mm Node: 727345

Fig. 3.15 Resultant displacement contour for FEA load case 2

Fig. 3.16 Manufacturing the cruciform

Fig. 3.17 Initial assembly of the crosshead within the Smart Structures Laboratory

reduce stress concentrations in zones such as connections and corners. The introduction of the curved stiffener plates connecting each arm of the crosshead successfully reduced the local stresses at each corner of the crosshead from 325 to 175 MPa. The strength of the welds was also carefully examined and designed to contain a tensile strength greater than 450 MPa. The maximum vertical displacement recorded under a vertical peak actuator load of 4 MN was 2.7 mm. This value confirmed that the stiffness of the crosshead was sufficient for the intended use of the facility and that the displacements of the cruciform under load were within an acceptable level of tolerance. Figure 3.15 illustrates the resultant displacement contour for FE analysis load case 2. Figures 3.16 and 3.17 show the cruciform under manufacture and its final shape in the Smart Structures Laboratory.

3.3 MAST Actuator Assembly and 6-DOF Control System

The actuator configuration and control system of the MAST have been designed for accurate and stable simulation of complex 6-DOF boundary effects. Table 3.3 summarizes the actuator specifications of the system, and Fig. 3.18 schematically shows the control point and actuator arrangement. In this configuration, two sets of actuator pairs with strokes of ± 250 mm provide lateral loads up to ± 1 MN in the

Table 3.3 Summary of actuator specifications

MAST actuators capacity				
Actuator	Vertical	Horizontal	Auxiliary	
Model	MTS 244.51	MTS 244.41	2 (MN)	(Qty. 1)
Quantity	4 (Z_1, Z_2, Z_3, Z_4)	4 (X_1, X_2, Y_3, Y_4)	250 (kN)	(Qty. 4)
Force stall capacity	$\pm$ 1000 (kN)	$\pm$ 500 (kN)	100 (kN)	(Qty. 3)
Static	$\pm$ 250 (mm)	$\pm$ 250 (mm)	25 (kN)	(Qty. 3)
Servo-valve flow	114 (lpm)	57 (lpm)	10 (kN)	(Qty. 1)

MAST DOFs capacity (non-concurrent)			
DOF	Load	Deformation	Specimen dimension
X (Lateral)	1 (MN)	$\pm$ 250 (mm)	3.00 (m)
Y (Longitudinal)	1 (MN)	$\pm$ 250 (mm)	3.00 (m)
Z (Axial/Vertical)	4 (MN)	$\pm$ 250 (mm)	3.25 (m)
Rx (Bending/Roll)	4.5 (MN.m)	$\pm$ 7 (degree)	
Ry (Bending/Pitch)	4.5 (MN.m)	$\pm$ 7 (degree)	
Rz (Torsion/Yaw)	3.5 (MN.m)	$\pm$ 7 (degree)	

orthogonal directions. These actuator pairs are secured to the L-shaped strong wall (Fig. 3.18a). Four $\pm$ 1 MN vertical actuators, capable of applying a total force of $\pm$ 4 MN with strokes of $\pm$ 250 mm, connect the crosshead and the strong floor (Fig. 3.18b). The actuator positions and the control point in 3-D space are also illustrated in Fig. 3.18c. In addition, Figures A–4 and A–5 in Appendix provide the specifications for vertical and horizontal actuators, respectively. In addition to MAST actuators, a set of auxiliary actuators are also available to be used for additional loading configurations on the specimen. The actuator system specifications are summarized in Table 3.3. The vertical actuators have twice as much flow capacity as the horizontal actuators so that all actuators are mechanically limited to the same peak saturation velocity to simplify control system tuning.

Figures 3.19 and 3.20, respectively, illustrate the final configuration of the actuator assembly for the MAST system and the hydraulic power unit (Model 505.180) in the Smart Structures Laboratory.

The movement of the MAST crosshead is governed by the collective movement of all eight actuators. To achieve the desired crosshead linear and angular displacements, the actuators are synchronously controlled using the MTS DOF control concept (Thoen 2013). This concept allows the user to control the system motion in a coordinate domain most natural to the test. With multiple actuators positioning the crosshead, it is impractical to control the system by individually controlling each actuator. Therefore, the MTS controller provides simultaneous control of the 6-DOF movements of the crosshead with respect to the control point, where it is attached to the specimen (see Fig. 3.18b). In DOF control, the feedbacks for each loop are determined by summing together all individual actuator feedbacks that contribute to that specific DOF, and each actuator drive-signal is determined by summing together all individual DOF error signals that are affected by that actuator.

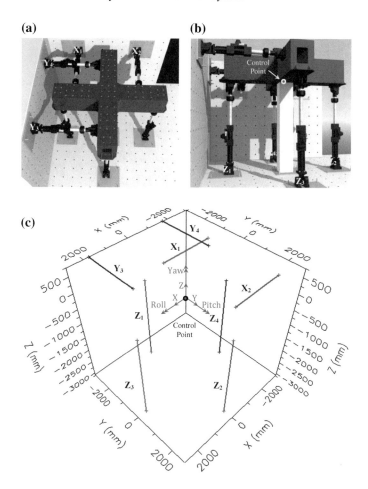

Fig. 3.18 Actuator assembly and positions: **a** Actuator assembly: plan view, **b** actuator assembly: side view, **c** actuator positions and the control point in 3-D space

Therefore, the control loop is closed around the DOFs and not around individual actuators.

The MAST system uses a generic kinematic transformation package that transforms the eight actuator displacements from the actuator domain into the 6-DOF coordinate system. The DOF command and control coordinates are defined in Cartesian space and referenced to the specimen initial control point and axes. The Cartesian DOF coordinate vector is written in vector form as:

$$u = [x, y, z, \theta_x, \theta_y, \theta_z]^T \tag{3.1}$$

Fig. 3.19 MAST system

Fig. 3.20 Hydraulic power unit, Model 505.180 (600 lpm)

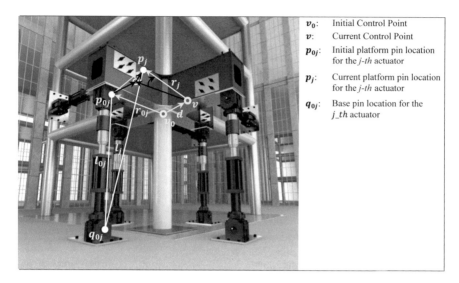

v_0:	Initial Control Point
v:	Current Control Point
p_{0j}:	Initial platform pin location for the j-th actuator
p_j:	Current platform pin location for the j-th actuator
q_{0j}:	Base pin location for the j_th actuator

Fig. 3.21 Actuator kinematics

The actuator displacements needed to generate the desired 6-DOF movement are defined as the lengths between the spherical joints at the ends of each actuator:

$$l = [l_1, l_2, l_3, l_4, l_5, l_6, l_7, l_8]^T \qquad (3.2)$$

The 3D notation used to define the actuator spherical joint locations in vector form is illustrated in Fig. 3.21. The fixed-pin location of the actuators to the strong floor/wall is defined as the point, q_{0j}. The moving-pin location of the actuators attached to the crosshead is defined as the point, p. The 3-D definition of these 16 points (i.e., the two spherical joint locations of the eight actuators) and the initial control point are the key geometric parameters for the kinematic relationship between the DOF and actuator coordinates. The initial actuator length (i.e., the length of the actuator when the control point is at origin) for the j-th actuator can then be written in the following form:

$$l_{0j} = |p_{0j} - q_{0j}| \qquad (3.3)$$

The translational DOFs, $d = [x, y, z]^T$, in u can be written in terms of the initial and current control points as follows:

$$v = |v_0 + d| \qquad (3.4)$$

In addition, r_{0j} is defined as a vector from the initial control point to the initial crosshead pin location of the j-th actuator.

$$r_{0j} = p_{0j} - v_0 \tag{3.5}$$

The rotational displacements $(\theta_x, \theta_y, \theta_z)$ result in a pure rotation of r_{0j}.

$$r_j = \psi r_{0j} \tag{3.6}$$

where the rotational matrix ψ follows the roll–pitch–yaw rotational convention and is given by:

$$\psi = \begin{bmatrix} \cos\theta_z & -\sin\theta_z & 0 \\ \sin\theta_z & \cos\theta_z & 0 \\ 0 & 0 & 1 \end{bmatrix} \begin{bmatrix} \cos\theta_y & 0 & \sin\theta_y \\ 0 & \cos\theta_y & 0 \\ -\sin\theta_y & 0 & 1 \end{bmatrix} \begin{bmatrix} 1 & 0 & 0 \\ 0 & \cos\theta_x & -\sin\theta_x \\ 0 & \sin\theta_x & \cos\theta_x \end{bmatrix} \tag{3.7}$$

The current crosshead pin location, p_j, for the j-th actuator due to the motion, u, is the sum of the translational displacement vector at the control point, d, and rotated vector, r_j:

$$\begin{aligned} p_j &= v + r_j \\ &= v_0 + d + \psi_{0j} \end{aligned} \tag{3.8}$$

Finally, the current actuator length for the j-th actuator, l_j, can be written as follows:

$$\begin{aligned} l_j &= \left| p_j - q_{oj} \right| \\ &= \left| v_0 + d + \psi r_{0j} - q_{0j} \right| \\ &= \left| p_{0j} - r_{0j} + d + \psi r_{0j} - q_{0j} \right| \\ &= \left| d - (I - \psi) r_{0j} + p_{0j} - q_{0j} \right| \end{aligned} \tag{3.9}$$

Equation 3.9 is a kinematic relationship between the j-th actuator length, l_j, and the global Cartesian DOF coordinate vector, u, which is integrated into the controller. Note that the discussed actuator kinematics is a simplification of the kinematic transformations in the control system. The MAST system has further flexibilities to relocate/tare/offset the initial control point, v_0, from the crosshead center point and also relocate/reorient the center of rotation (COR) defined as the origin of the Cartesian DOF coordinate system.

The initial control point, v_0, is initially placed at the crosshead center point when it is completely horizontal with all four Z actuators absolutely vertical and the two X and two Y actuators absolutely horizontal. This is very close to the point where all actuators are at their mid-travel. The ability to offset the initial control point allows for variations in specimen height and errors in initial specimen positioning and alignment. It also allows the crosshead to tilt like a spherical joint (pin) to adjust for a specimen when the top and bottom surfaces are not exactly parallel. The offsetting allows for the test command vector, u, to be the top center of the test specimen or any other point if desired.

The center of rotation (COR) is typically placed at the initial control point, v_0, with all DOF displacements and forces parallel to the strong walls/floor. However, using the COR vector $= \left[\bar{x}, \bar{y}, \bar{z}, \bar{\theta}_x, \bar{\theta}_y, \bar{\theta}_z \right]^T$ featured in the control system, the COR can be relocated and/or reoriented by assigning the desired values to $\left[\bar{x}, \bar{y}, \bar{z} \right]$ and/or $\left[\bar{\theta}_x, \bar{\theta}_y, \bar{\theta}_z \right]$, respectively. For instance, by setting COR vector $= [0, 0, 0, 0, 0, 45°]^T$, the COR is not relocated as $\left[\bar{x}, \bar{y}, \bar{z} \right]$ is set to zero; however, COR is rotated around the Z-axis by $\bar{\theta}_z = 45°$. This increases the lateral deformation capacity of the system from ± 250 mm in the initial X and Y axis to ± 350 mm ($= \pm 250$ mm $\times \sqrt{2}$) in the 45° rotated X and Y axis. The other two angles, $\bar{\theta}_x$ and $\bar{\theta}_y$, can be adjusted so that the forces are applied normal to the top surface of the test specimen.

The MAST system also features mixed-mode control, allowing users to specify the deformation or load required for the desired direction of loading to the test specimen. Figure 3.22 schematically shows the MAST control system that includes mixed-mode motion control and the force-balance control. As illustrated in the figure, the measured actuator force and displacements are transformed into the load/deformation vector in the DOF coordinate system and compared with the target load/deformation commands. The controller sees the difference as the error and minimizes that by calculating the updated DOF command signals. Using the geometric transformation, the DOF commands are converted to the actuator servo-valve commands, which are executed simultaneously. The measured displacement and force in the actuator's domain are then transformed back into the DOF system to provide the feedback signals for the controller.

The MAST control system also uses a force-balance control to manage the redundancy in the actuation system. Since the MAST system has eight actuators operating to control 6-DOFs, it is considered an over-constrained system. In such a system, the actuator displacements are constrained but actuator forces are not, and thus, an infinite number of actuator force states can satisfy a given position profile in the DOF system. As the crosshead is designed to have a very high stiffness, tiny offsets or inconsistencies in actuator positions can generate large distortion forces, which could appear as very large internal vertical and horizontal shear forces within the crosshead. The force imbalances can seriously limit the performance of the system when applying large forces to the specimen, as some portion of the actuators' force will be used against each other and consequently not be available to apply to the test specimen. Force-balance compensation corrects the actuator servo-valve commands by minimizing the magnitude sum of force errors and therefore ensures that the force is distributed equally among all driving actuators.

The MAST system uses external redundant high-precision draw-wire absolute encoders, in addition to the actuators' linear variable differential transformer

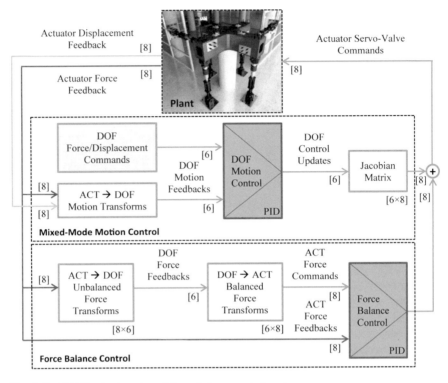

Fig. 3.22 MAST mixed-mode and force-balance control system

(LVDT) position sensors, in order to measure the actuator length over the range of control. The string encoders (see Fig. 3.23) have the resolution of 25 μ and can be used to calibrate the linearization tables for the LVDTs or directly for displacement feedback in the controller.

Fig. 3.23 High-precision draw-wire absolute encoders

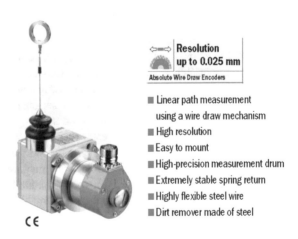

Resolution up to 0.025 mm
Absolute Wire Draw Encoders

■ Linear path measurement using a wire draw mechanism
■ High resolution
■ Easy to mount
■ High-precision measurement drum
■ Extremely stable spring return
■ Highly flexible steel wire
■ Dirt remover made of steel

CE

3.4 Hybrid Simulation Architecture

The hybrid simulation control system consists of a three-loop architecture (Stojadinovic et al. 2006) as follows:

1. The innermost servo-control loop contains the MTS FlexTest 100 controller that sends the actuator servo-valve commands while reading back measured displacements/forces. The displacements are measured from both the actuators' LVDTs and the high-precision draw-wire absolute encoders.
2. The middle loop runs the predictor-corrector algorithms for actuator command generation on the xPC-Target digital signal processor (DSP) and delivers the displacement/force commands to the FlexTest controller in real time through the shared memory SCRAMNet (Systran 2004).
3. The outer integrator loop runs on the xPC-Host and includes OpenSees, OpenFresco and MATLAB/Simulink, which can communicate with the xPC-Target through the TCP/IP network.

The FlexTest 100 controller runs the MTS 793 DOF software along with the MTS Test Suite and the MTS Computer Simulation Package that provides the fiber optic SCRAMNet interface. At present, the eight absolute encoders are connected in parallel with each actuator. However, it is also possible to relocate these closer to the test specimen to provide 6-DOF control at the specimen in a similar way to using strain control in a SDOF test machine. In this manner, any errors due to crosshead wall and floor deflections can be canceled out.

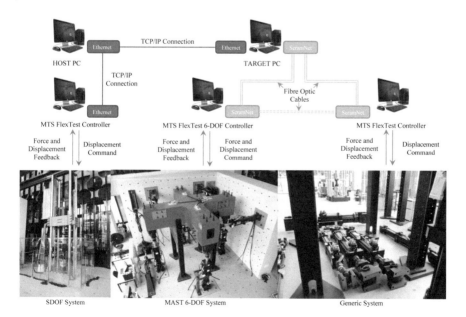

Fig. 3.24 Hybrid simulation systems in the Smart Structures Laboratory at Swinburne

It should be mentioned that the hybrid simulation architecture at Swinburne is not limited to the MAST system. A generic configuration of actuators, especially for the case of experimental testing at system level, could also be integrated into hybrid simulation. Further, a SDOF 1MN universal testing machine is available that is suitable for proof-of-concept tests in hybrid simulation studies. Figure 3.24 illustrates an overview of hybrid simulation architecture at Swinburne.

References

ANSYS Inc. (2012). Engineering Simulation & 3-D Design Software, ANSYS. Canonsburg, PA, U.S.

French, C. E., Schultz, A. E., Hajjar, J. F., Shield, C. K., Ernie, D. W., Dexter, R. J., et al. (2004). Multi-axial subassemblage testing (mast) system: Description and capabilities. In *13th World Conference on Earthquake Engineering*. Vancouver, Canada.

Hashemi, M. J., Al-Mahaidi, R., Kalfat, R., & Burnett, G. (2015). Development and validation of multi-axis substructure testing system for full-scale experiments. *Australian Journal of Structural Engineering*. doi:10.1080/13287982.2015.1092692.

Hofmann. (2013). Hofmann Engineering Pty. Ltd. Perth, Australia.

Kim, S. J., Holub, C. J., & Elnashai, A. S. (2011). Experimental investigation of the behavior of RC bridge piers subjected to horizontal and vertical earthquake motion. *Engineering Structures, 33*(7), 2221–2235.

Mahmoud, H. N., Elnashai, A. S., Spencer, B. F., Kwon, O. S., & Bennier, D. J. (2013). Hybrid simulation for earthquake response of semirigid partial-strength steel frames. *Journal of Structural Engineering (ASCE), 139*(7), 1134–1148.

McKenna, F. (2011). Opensees: A framework for earthquake engineering simulation. *Computing in Science & Engineering, 13*(4), 58–66.

MTS Systems Corporation. (2014). Manufacturer of Testing System and Sensing Solutions. Eden Prairie, Minnesota, USA.

Schellenberg, A. H., Mahin, S. A., & Fenves, G. L. (2009). *Advanced implementation of hybrid simulation*. Berkeley, U.S.: Pacific Earthquake Engineering Research Center, University of California.

SICK AG. (2014). Manufacturer of Sensors and Sensor Solutions. Waldkirch, Germany.

Stojadinovic, B., Mosqueda, G., & Mahin, S. A. (2006). Event-driven control system for geographically distributed hybrid simulation. *Journal of Structural Engineering, 132*(1), 68–77.

Systran, C. (2004). *The SCRAMNet + Network (Shared Common RAM Network)*. Dayton, U.S.

The MathWorks Inc. (2014). *MATLAB R2014b*. U.S.: Massachusetts.

Thoen, B. (2013). *Generic kinematic transforms package*. Minneapolis, U.S.: MTS Systems Corporation.

Waterman Group plc. (2010). Consulting Engineers and Environmental Scientists. London, UK.

Chapter 4
Application of the MAST System for Collapse Experiments

Abstract This chapter presents the results of a range of experiments, including switched/mixed load/deformation mode quasi-static cyclic and hybrid simulation tests to highlight the unique and powerful capabilities of the MAST system, specifically for the assessment and mitigation of the collapse risk of structures.

Keywords Switched and mixed mode control · Quasi-static cyclic · Hybrid simulation · Collapse experiments

4.1 Introduction

One of the main goals of structural/earthquake engineering is to improve the resilience and performance of structures to withstand collapse due to extreme events. Recent devastating events including a number of large magnitude earthquakes around the world (e.g., Northridge 1994, Kobe 1995, Chile 2010, East Japan and Christchurch 2011) demonstrated that their intensity could reach to more than two times the design level according to the regional code provisions. In this context, it is becoming increasingly important to quantify the reserve capacity of structures against extreme events beyond the design level to the levels approaching collapse. Although there have been many advancements in the mathematical models employed in finite-element methods, many of these analytical models are calibrated using experimental observations. Therefore, experimental research remains critical toward better understanding and predicting the response of structures. However, there are also challenges in conducting laboratory tests for a number of reasons.

Firstly, actions on structures during extreme events such as earthquakes are generally multi-directional and continuously varying, due to the time-dependent nature of the input motion. For instance, variations of the axial loads during a seismic excitation may influence the response of the vertical structural components

The original version of this chapter was revised: See the "Chapter Note" section at the end of this chapter for details. The erratum to this chapter is available at https://doi.org/10.1007/978-981-10-5867-7_6

(e.g., bridge piers and building columns) since the response of such elements when combined with flexural, shear and torsional actions may differ from the cases when they are not subjected to the same axial load changes. Simulation of such highly coupled multi-directional loading conditions using conventional structural testing methods can be expensive, time-consuming and difficult to achieve. As a result, advanced and innovative experimental techniques and control strategies are under development by researchers (Nakata 2007; Wang et al. 2012; Hashemi et al. 2014; Hashemi and Mosqueda 2014).

Secondly, the experiments should be conducted large or full scale to accurately capture the local behavior of the elements. Certain types of local behavior such as bond and shear in reinforced concrete (RC) members, crack propagation, welding effects and local buckling in steel structures are well known to have size effects. Conducting large-scale experiments, however, may not be feasible due to the limited resources available in many laboratories including the number and capability of available actuators, the dimensions and load capacity of the reaction systems, difficulties in the actuator assemblies and testing configuration to reliably simulate the boundary conditions (Hashemi et al. 2016). Consequently, the specimen may be tested small scale or under uni/biaxial loading configurations, which do not necessarily represent the actual action or demand on the structural elements and the corresponding nonlinear response of the prototype system.

Finally, conducting multi-directional loading including axial load effects requires switched-/mixed-mode control strategy. The application of axial loads has been mainly considered by researchers using a combination of force-control actuators in the vertical direction that are decoupled from displacement-control actuators in the lateral direction of the specimen (Lynn et al. 1996; Pan et al. 2005; Del Carpio Ramos et al. 2015). In those tests, independent of lateral actuators, only the vertical force-control actuators apply the axial forces, while under large deformations, lateral actuators have a force component in the vertical direction that needs to be accounted for. Therefore, versatile and generally applicable switched-/mixed-mode control algorithms are required to take into account instantaneous three-dimensional coupling in the control systems.

In this chapter, the unique and powerful capabilities of the MAST system are presented in application for collapse assessment of reinforced concrete (RC) building components. The first experiment is a switched-mode quasi-static test of the RC wall subjected to pure axial demands through displacement-controlled tension and force-controlled compression loads. The next two experiments are conducted on two identical RC columns that are tested through mixed-mode quasi-static and hybrid simulation tests through collapse. In the quasi-static test, the specimen is subjected to constant axial load combined with bidirectional deformation reversal with increasing amplitude that follows a hexagonal orbital pattern. In hybrid simulation, the RC column serves as the first-story corner-column of a half-scale symmetrical 5 × 5 bay 5-story RC ordinary moment frame building that is subjected to bidirectional sequential ground motions with increasing intensities. Finally, to assess the effectiveness of the carbon fiber-reinforced polymer (CFRP) on rehabilitation of RC building columns, the earthquake-damaged RC column was repaired and retested under the same loading condition in hybrid simulation. A simplified collapse risk assessment study was then conducted to compare the response of the RC columns obtained from the quasi-static and hybrid simulation tests as well as the responses of the initial and repaired RC columns.

4.2   Switched-Mode Quasi-Static Test

In the first experiment, the MAST system was used to assess the performance of an RC wall in a building that collapsed during the 2011 Christchurch earthquake. The building was a 5-story RC structure with a lateral load-resisting system comprised of RC walls forming a large shear core, centrally located in the floor plan. The collapse mechanism of the building was hypothesized to be an out-of-plane buckling instability of one of the RC core walls.

Due to the configuration of doorways and wall openings in the shear core, one of the walls at the ground floor had no adjacent return segments. It was theorized that this resulted in the wall being subjected to cyclic tension-compression axial loads due to the overturning moments on the shear core during the earthquake. Large cyclic axial loads eventually caused high enough tensile strains on the wall that it became extremely susceptible to out-of-plane buckling instabilities on the reversed cycle, where the wall was subjected to axial compression.

To simulate the out-of-plane buckling of the critical wall, a 3:4 scale model was constructed with a cross section of 1050 mm × 148 mm and a height of 2680 mm. The wall had a single central layer of longitudinal and transverse reinforcement, with, respectively, 0.25 and 0.35% reinforcement ratios. This resulted in two N16 (normal ductility with 16 mm diameter) longitudinal bars and 14-mm-diameter transverse bars placed at 300 mm centers. The reinforcement had a yield stress of 570 MPa and an ultimate stress of 670 MPa. The wall was cast using a Grade 40 concrete mix (i.e., a 40 MPa characteristic compressive strength at 28 days). Despite the concrete being a standard Grade 40 mix, the mean compressive strength on the test day, determined from 10 concrete cylinders with 100 mm diameter, was approximately 65 MPa.

To determine the approximate loading profile of the axial load on the wall, a finite-element model of the full building was built and subjected to the recorded ground motions. The loading profile was applied to the test specimen using switched-mode control. In this method, the applied loading automatically switches between force-controlled and displacement-controlled loading protocols. When the test specimen is being 'pulled up' in tension, the concrete cracks and the reinforcement yields and undergoes plastic deformation. Therefore, displacement-controlled loading is suitable to load the specimen up to the required tension displacement. When the test specimen is being 'pushed down' in compression, the concrete undergoes very small displacement increments for the associated amount of force, which sometimes are beyond the resolution of the measurement system. Therefore, force-controlled loading should be implemented to accurately reach to

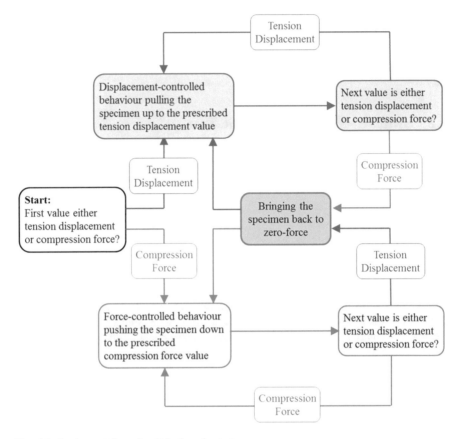

Fig. 4.1 Implementation of switched-mode strategy

the desired level of compressive load. The loading protocol procedure is graphically illustrated in Fig. 4.1. Note that, when switching back and forth between force and displacement, the specimen is brought back to zero force to avoid any control instabilities.

Figure 4.2a shows the wall tested under the MAST system along the Z axis, while the remaining 5 DOFs (i.e., X, Y, Rx, Ry and Rz) are commanded to zero deformation to allow for pure application of axial load on the specimen. Figure 4.2b shows the out-of-plane failure of the RC wall occurred at the base. The command and measured signals are also presented in Fig. 4.3. The switching points can be seen in force time history, where the applied load changes from tension to compression and vice versa. Also, it can be seen that the measured displacement and force signals precisely matched the respective command signals in tension and compression. In terms of the specimen response, it is observed that the fracture of a longitudinal rebar (at ∼ 1630 s) caused a sudden force reduction in tension, which was followed by the total failure of the specimen in compression.

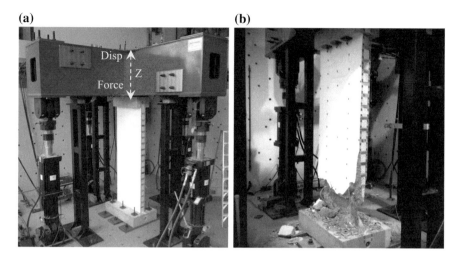

Fig. 4.2 Switched-mode quasi-static test of the RC wall: **a** RC wall subjected to displacement in tension and force in compression along the Z axis, **b** Out-of-plane failure of the RC wall

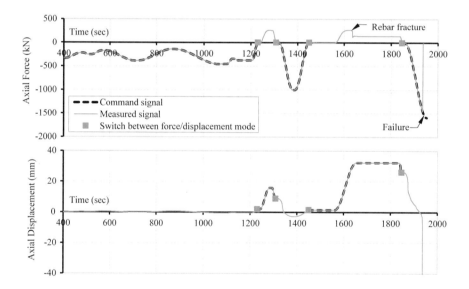

Fig. 4.3 Comparison of command/measured force/displacement signals

4.3 Mixed-Mode Quasi-Static Test

The second experiment is a three-dimensional mixed-mode quasi-static cyclic test conducted on a large-scale limited-ductility RC column. Figure 4.4a and b, respectively, shows the design details of the column and the 6-DOF axes of

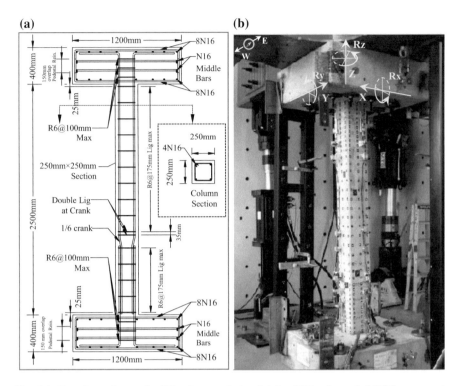

Fig. 4.4 Experimental setup for RC column: **a** design details of RC column, **b** 6-DOF movements of RC column

Table 4.1 Material properties of the RC column

Concrete		Steel	
ε_c	0.002	ε_y	0.0035
f_c'	35.1 (MPa)	f_y	633.95 (MPa)
ε_{cu}	0.0063	ε_u	0.073
f_u'	0.0	f_u	712.0 (MPa)

crosshead movements. The specimen is attached to the strong floor from the base and to the crosshead from the top through rigid concrete pedestals. The RC column is 2.5 m high, has a square 250 mm × 250 mm cross section and is reinforced with four longitudinal bars of N16 and tied with R6 stirrups spaced at 175 mm with 30 mm cover thickness. The material properties of the specimen, obtained from laboratory tests, are also presented in Table 4.1.

The loading protocol consists of simultaneously applying a constant gravity load, equal to 8% of ultimate compressive load capacity in force control, while imposing bidirectional lateral deformation reversals in displacement control that follows the hexagonal orbital pattern suggested in FEMA 461 (Federal Emergency Management Agency 2007), as shown in Fig. 4.5. The sequence of loading in QS

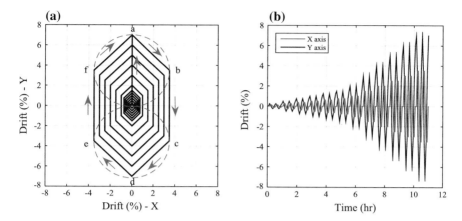

Fig. 4.5 QS loading protocol: **a** hexagonal orbital pattern for bidirectional lateral deformation reversals, **b** drift time histories in X and Y axes

testing started with applying the gravity load on the specimen along Z axis. The specimen was then pushed to the initial uniaxial drift ratio toward point 'a,' followed by the orbital pattern 'a-b-c-d-e-f-a.' The reversal from point 'a' accompanies an orthogonal drift at points 'b' and 'c' equal to one-half the maximum drift ratios at points 'a' and 'd.' The entire loading cycle was then repeated at the same amplitude. Once the specimen reached point 'a' for the second time, the amplitude value for the next two cycles was increased and the next two biaxial load cycles were applied on the specimen. The process continued until the failure of the specimen. The remaining DOF axes (roll, pitch and yaw) were controlled in zero angle forming a double-curvature deformation of the column.

The results of the QS test, including the hysteretic behavior of the RC column in X and Y axes, axial load time history in Z axis, biaxial lateral drifts in X and Y axes and biaxial bending moments in Rx and Ry axes are presented in Fig. 4.6. The force relaxations observed in the hysteresis were due to pausing of the test in order to collect photogrammetry data at peak deformations in the X axis. The failure of the specimen occurred when the specimen was subjected to the maximum of 7.0 and 3.5% drift ratios in Y and X axes, respectively. These are large drifts for a limited ductile column, but effective of the relatively low axial loads applied to the column (Wibowo et al. 2014).

4.4 Mixed-Mode Hybrid Simulation Test

The third experiment is a three-dimensional mixed-mode hybrid simulation using an identical RC column to the one previously tested in the quasi-static cyclic experiment. For this purpose, a half-scale symmetrical 5-story (height of first story $h_1 = 2.5$ m, height of other stories $h_{typ} = 2.0$ m) 5×5 bay (column spacing

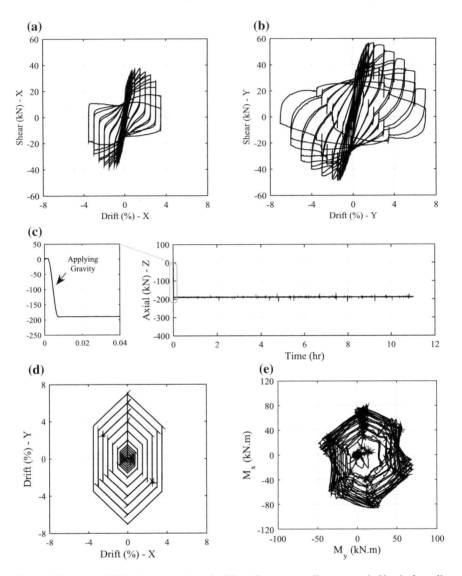

Fig. 4.6 Response of RC column specimen in QS cyclic test: **a** cyclic response in X axis, **b** cyclic response in Y axis, **c** axial load time history, **d** biaxial lateral drifts, **e** biaxial moment interactions

b = 4.2 m) RC ordinary moment frame building is selected as the hybrid model. The physical specimen serves as the first-story corner-column of the building, considered as the critical element of the structure. The rest of the structural elements, inertial and damping forces, gravity and dynamic loads and second-order effects are modeled numerically in the computer. Figure 4.7 illustrates the components of hybrid simulation including numerical and experimental substructures.

(a)   **(b)**

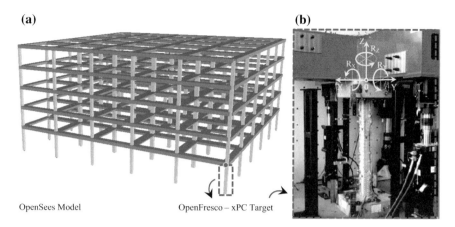

OpenSees Model   OpenFresco – xPC Target

Fig. 4.7 Hybrid simulation substructures: **a** numerical substructure, **b** experimental substructure

(a)   **(b)**

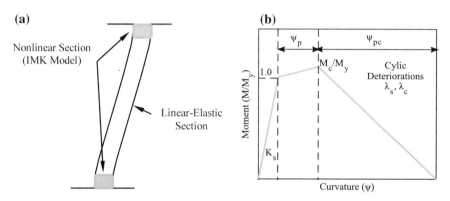

Fig. 4.8 Nonlinear analytical model for beam–column elements: **a** beam-with-hinges model, **b** modified Ibarra–Medina–Krawinkler (IMK) model

The structure's beams and columns were modeled using beam-with-hinges elements, where the nonlinear behavior is assumed to occur within a finite length at both ends based on the distributed plasticity concept (Scott and Fenves 2006) (Fig. 4.8a). The plasticity model follows a peak-ordinated hysteresis response based on the modified Ibarra–Medina–Krawinkler (IMK) deterioration model of flexural behavior (Ibarra et al. 2005; Zhong 2005).This model was chosen because it is capable of capturing the important modes of deterioration that participate in sidesway collapse of RC frames. The model requires the specification of a range of parameters to control the tri-linear monotonic backbone curve and different modes of cyclic deteriorations. As shown in Fig. 4.8b, these parameters include M_y, M_c/M_y, ψ_p, ψ_{pc} and λ that, respectively, represent yielding moment, a measure of maximum moment capacity, plastic curvature capacity, post-capping curvature capacity and cyclic deterioration. The model captures four modes of cyclic

Table 4.2 IMK model parameters for RC frame structure

Story no.	Element location	M_c/M_y	M_y(kN.m)	EI/EI_g	ψ_p	ψ_{pc}	λ_S, λ_C
Story 1 Columns	Corner	1.20	114	0.52	0.12	9.00	14.0
	Exterior	1.20	128	0.60	0.10	7.00	10.5
	Interior	1.18	467	0.48	0.08	8.70	28.0
Story 2 Columns	Corner	1.18	111	0.41	0.13	9.20	15.4
	Exterior	1.20	124	0.48	0.11	7.61	11.9
	Interior	1.18	464	0.38	0.08	9.13	30.1
Story 3 Columns	Corner	1.18	109	0.39	0.13	9.76	16.1
	Exterior	1.20	117	0.44	0.12	8.12	13.3
	Interior	1.18	450	0.35	0.09	10.0	32.9
Story 4 Columns	Corner	1.18	106	0.37	0.13	9.96	16.8
	Exterior	1.18	112	0.41	0.13	9.18	15.4
	Interior	1.18	435	0.35	0.09	10.2	35.0
Story 5 Columns	Corner	1.18	102	0.36	0.14	10.4	18.2
	Exterior	1.18	106	0.37	0.14	9.71	16.8
	Interior	1.18	419	0.35	0.09	10.6	37.8
Beams	Exterior end span	1.195	160	0.35	0.11	8.99	29.6
	Exterior span		152	0.35	0.11	8.83	31.1
	Interior end span		247	0.35	0.16	8.99	46.5
	Interior span		246	0.35	0.13	9.02	38.2

deterioration, including strength deterioration of the hardening region (λ_S), strength deterioration of the post-peak softening region (λ_C), accelerated reloading stiffness deterioration (λ_A) and unloading stiffness deterioration (λ_K). Based on the studies by Haselton et al. (2008), the strength deterioration of hardening region and the post-capping strength deterioration were assumed to be equal in the case study, while accelerated reloading and unloading stiffness deterioration were ignored. This reduced the calibration of cyclic deteriorations to one parameter. Table 4.2 presents the IMK parameters for beam and column elements.

After developing the numerical model, the elastic fundamental period of vibration and the corresponding first mode shape were obtained through eigenvalue analysis. A nonlinear static pushover analysis was then performed with the lateral force distribution proportional to the fundamental mode of vibration and with the consideration of second-order $P - \Delta$ effects. Figure 4.9 presents the results of the pushover analysis that show most of the energy dissipation occurs in the lower two stories.

For the HS test, the two horizontal components of the 1979 Imperial Valley earthquake ground motions recorded at El Centro station with peak ground acceleration of 0.15 g were used. Figure 4.10 shows the acceleration, displacement and acceleration–displacement response spectra for the two ground-motion components. Based on incremental dynamic analysis, four levels of intensities were considered

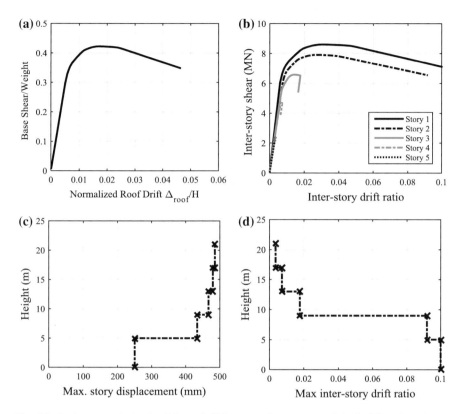

Fig. 4.9 Pushover analysis of RC frame building: **a** pushover curve of the building, **b** pushover curve of individual stories, **c** maximum floor displacements, **d** maximum inter-story drift ratios

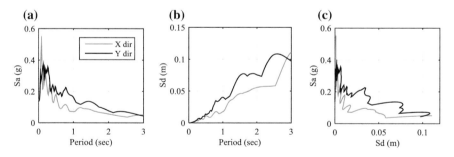

Fig. 4.10 Response spectra for the two horizontal components of the 1979 Imperial Valley earthquake ground motions (recorded at El Centro station) used in HS test: **a** acceleration response spectra, **b** displacement response spectra, **c** acceleration–displacement response spectra

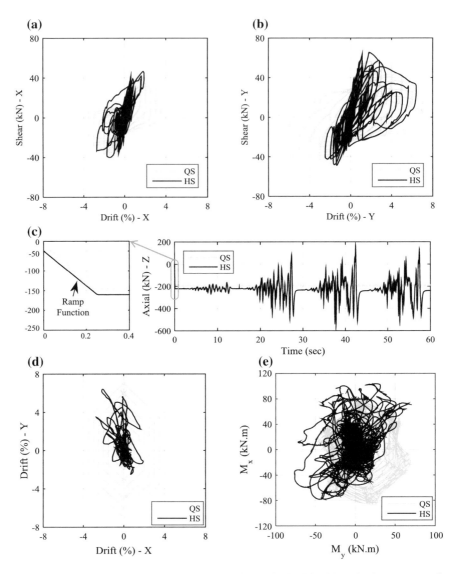

Fig. 4.11 Responses of the RC column and applied axial load in QS and HS tests: **a** cyclic response in X axis, **b** cyclic response in Y axis, **c** axial load time history, **d** biaxial lateral drifts, **e** biaxial moment interactions

to capture the full range of structural response from linear elastic range to collapse. The selected scale factors were 0.6, 4.0, 8.0 and 9.0, which pushed the structure to nearly 0.25% (elastic), 2, 4 and 6% inter-story drift ratios, respectively.

Prior to conducting the actual HS test with the physical subassembly in the laboratory, a series of FE-coupled numerical simulations (Schellenberg et al. 2009) was conducted to evaluate the integration scheme parameters for the actual experiments. Accordingly, Generalized Alpha-OS (Schellenberg et al. 2009) was used as the integration scheme, and the integration time step was optimized to preserve the accuracy and stability of the simulation, while allowing the completion of the entire test during the regular operation time of the laboratory. Five percentage Rayleigh damping was specified to the first and third modes of vibration, corresponding to the primary translational modes in X and Y directions. Additional damping was also assigned to free vibration time intervals between the forced vibrations in order to quickly bring the structure to rest.

The hybrid simulation started with applying the gravity load on the specimen, using a ramp function, followed by sequential ground motions. The entire sequence of loading was performed and automated using OpenSees. Considering the 117-ms delay in the hydraulic system, 500 ms was specified as the simulation time step in xPC-Target predictor-corrector to provide sufficient time for integration computation, communication, actuator motions and data acquisition. This scaled the 60 s of sequential ground motions to 6 h in laboratory time. Similar to the QS test, the rotational axes (roll, pitch and yaw) were controlled in zero angle, forming a double-curvature deformation of the column.

Figure 4.11 compares the responses of RC column including the hysteretic behavior of the RC column in X and Y axes, axial load time history in Z axis, biaxial lateral drifts in X and Y axes and biaxial bending moments in Rx and Ry axes for the QS and the HS tests. The specimen was pushed to a maximum of 6.4 and 2.7% drift ratios in Y and X axes, respectively. The maximum time-varying axial load applied on the specimen was 553 kN in compression and 161 kN in tension. By comparing the hysteresis plots from the HS test, it can be seen that the column was damaged as the structure progressively moved in one direction, while in the QS test the pattern of damage was symmetrical due to load reversals in cyclic deformations. Figure 4.12 shows the flexural failure of columns for both tests by comparing the plastic hinges developed at the top and the base of the columns from different angles.

Test	Top - west view	Top - east view	Base - west view	Base - east view
QS				
HS				

Fig. 4.12 Comparison of plastic hinges in QS and HS tests

4.5   Comparison of Mixed-Mode Quasi-Static and Hybrid Simulation Tests

4.5.1   Numerical Model and Calibration

In order to investigate the influence of the selected experimental method on assessing the collapse risk, a comparative collapse fragility analysis for a sub-structure of the RC building was conducted using the results of QS and HS tests, respectively. As illustrated in Figure 4.13, the numerical model selected for incremental dynamic analysis includes only the first-story corner-column and the overhead mass portion of the upper five floors, which is equivalent to a single-degree-of-freedom (SDOF) system with a natural period of 0.6 s. This allows the study of the response of the critical element (i.e., the first-story corner-column) purely based on experimental results without the influence of other elements' responses.

The experimental results were used to calibrate the SDOF numerical model. As previously shown in Fig. 4.8, the moment-curvature behavior of the plastic zones follows the IMK hysteresis model. Although this model can generally simulate most of the important behaviors, including strength and stiffness degradation, effects such as the interaction between axial, flexure and shear failure cannot be captured. Accordingly, a unidirectional numerical model of the column was selected, and the hysteresis parameters of IMK model were calibrated to the response of the specimen in the main axis (i.e., the Y axis of the MAST system), along which it experienced maximum deformation. Consequently, the influence of axial loads and out-of-plane movements in the experiments were implicitly taken into account by using the calibrated numerical models. Note that the use of fiber-based plasticity models could be an alternative. However, only the most basic aspects such as material constitutive relationships are modeled, while the degradation parameters that have a significant impact on collapse behaviors are not included (Deierlein et al. 2010).

A close view of the hysteretic responses of the RC columns in the QS and HS tests is presented in Fig. 4.14. It can be clearly observed that the flexural strength, the capping point and post-capping negative tangent stiffness (in-cycle strength degradation) are significantly different. Accordingly, following the procedures

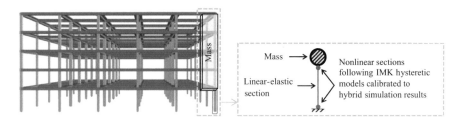

Fig. 4.13 Numerical substructure selected for collapse risk assessment

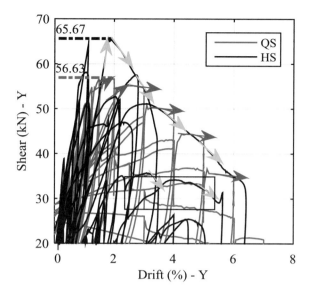

Fig. 4.14 Close view of the hysteretic responses in QS and HS tests

given in Haselton et al. (2008), the numerical SDOF model was calibrated to the QS and HS test results, with particular focus on precisely mimicking the plastic and post-capping deformation capacities as well as the cyclic deteriorations that are known to have important influence on collapse prediction (see Fig. 4.15). However, it is noted that some portions of unloading and reloading phases in the experiments, especially for the QS test, resulted in higher strength and larger hysteretic loops in the experiment compared to the more pinched behavior in calibrated models.

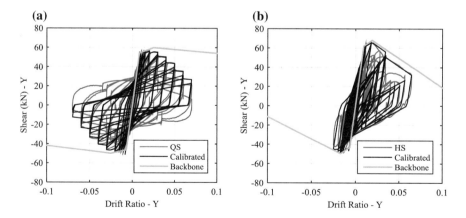

Fig. 4.15 Calibration of SDOF numerical models: **a** QS test, **b** HS test

Table 4.3 Comparison of IMK model parameters calibrated to QS and HS tests

Experiment method	M_c/M_y	M_y^+ (kN.m)	M_y^- (kN.m)	EI/EI_g	ψ_p^+	ψ_p^-	ψ_{pc}^+	ψ_{pc}^-	λ_S, λ_C
QS	1.12	68.0	58.0	0.258	0.14	0.18	6.0	5.0	6.0
HS	1.12	78.0	54.0	0.258	0.07	0.12	0.9	1.0	9.2

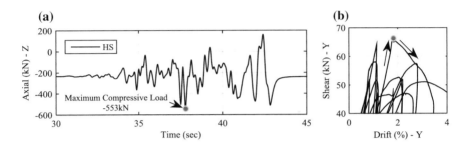

Fig. 4.16 Maximum in-cycle negative tangent stiffness corresponding to maximum compressive axial load in HS test: **a** axial time history, **b** hysteretic response

Table 4.3 compares the IMK model parameters for QS- and HS-based numerical models. The higher flexural strength and significant reduction in the drift capacity observed in the HS test may be due to the higher levels of axial load in this test, as previously addressed (Lynn et al. 1996; Nakamura and Yoshimura 2002; Sezen 2002). Figure 4.16 specifically shows that the rapid drop in shear strength occurred immediately after the maximum compressive axial load in the HS test, which is 2.9 times larger than the uniform axial load applied in the QS test. This clearly shows the significant impacts of the axial load level and its variation on the performance of RC structures and the ultimate drift capacity. Studies conducted by Wibowo et al. (2014) reported similar findings based on an experimental program performed to develop a generic backbone pushover curve for lightly reinforced concrete columns.

Another observation is that the specimen showed larger cyclic deteriorations in the QS test due to the application of many large cycles and load reversals to the specimen before failure. Note that the values assigned to the cyclic deterioration in the IMK model are inversely related to the level of deterioration.

4.5.2 Fragility Analysis

Incremental dynamic analyses (IDAs) were performed using the calibrated numerical model in order to capture a range of probable dynamic response behaviors due to record-to-record variability in ground-motion characteristics. For this purpose, three earthquake scenarios including M6.0R28, M6.5R40 and M7.0R90 (M and R stand for magnitude and source-site distance, respectively) were considered. A suite of 20 recorded ground motions was selected from the PEER database (PEER 2013) that are listed in Table 4.4 along with the values of peak ground acceleration (PGA) and spectral acceleration at the fundamental natural period of the numerical model $S_a(T_1)$

Table 4.4 List of input ground motions used in incremental dynamic analysis

No.	Record Name	Scenario	$PGA(g)$	$S_a(T_1)(g)$
1	WHITTIER 10/01/87 14:42 (USC STATION 90017)	M6.0 R28	0.0365	0.0798
2	CHICHI AFTERSHOCK 09/20/99 1757, TCU076, E (CWB)	M6.0 R28	0.0716	0.0719
3	CHICHI AFTERSHOCK 09/20/99 1757, TCU088, E (CWB)	M6.0 R28	0.0584	0.0118
4	CHICHI AFTERSHOCK 09/20/99 1757, TCU129, E (CWB)	M6.0 R28	0.0927	0.0366
5	CHICHI AFTERSHOCK 09/20/99 1803, CHY074, E (CWB)	M6.0 R28	0.0617	0.2227
6	SAN FERNANDO 02/09/71 14:00 (CDWR STATION 269)	M6.0 R28	0.1018	0.0580
7	NORTHRIDGE 1/17/94, 12:31 (CDMG STATION 24305)	M6.5 R40	0.0889	0.1654
8	NORTHRIDGE 1/17/94, 12:31 (CDMG STATION 24307)	M6.5 R40	0.0843	0.1935
9	NORTHRIDGE 1/17/94, 12:31 (CDMG STATION 24521)	M6.5 R40	0.0614	0.1485
10	NORTHRIDGE 1/17/94, 12:31 (CDMG STATION 24644)	M6.5 R40	0.0909	0.2497
11	CHICHI AFTERSHOCK 09/25/99 2352, CHY029, E (CWB)	M6.5 R40	0.2393	0.1636
12	CHICHI AFTERSHOCK 09/25/99 2352, CHY035, E (CWB)	M6.5 R40	0.1709	0.3447
13	CHICHI AFTERSHOCK 09/25/99 2352, HWA020, E (CWB)	M6.5 R40	0.0260	0.0425
14	CHICHI AFTERSHOCK 09/25/99 2352, HWA035, E (CWB)	M6.5 R40	0.0218	0.0520
15	CHICHI AFTERSHOCK 09/25/99 2352, HWA058, E (CWB)	M6.5 R40	0.0465	0.0570
16	CHICHI AFTERSHOCK 09/25/99 2352, TCU048, E (CWB)	M6.5 R40	0.0365	0.0798
17	CHICHI AFTERSHOCK 09/25/99 2352, TCU087, E (CWB)	M6.5 R40	0.0217	0.0399
18	CHICHI AFTERSHOCK 09/25/99 2352, TCU104, E (CWB)	M6.5 R40	0.0282	0.0733
19	CHICHI AFTERSHOCK 09/25/99 2352, TCU136, N (CWB)	M6.5 R40	0.0391	0.0258
20	LOMA PRIETA 10/18/89 00:05 (CDMG STATION 58043)	M7.0 R90	0.0713	0.1294

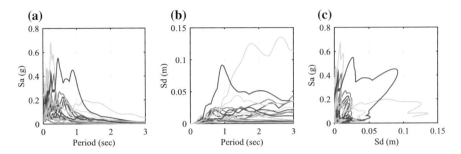

Fig. 4.17 Response spectra for ground motions used in IDA: **a** acceleration response spectra, **b** displacement response spectra, **c** acceleration–displacement response spectra

and $T_1 = 0.6$ s for the SDOF model. The response spectra of the input ground motions are also shown in Fig. 4.17.

Each unidirectional ground motion was individually applied to QS- and HS-based calibrated numerical models for the nonlinear simulation. The ground motions were increasingly scaled according to the value of $S_a(0.6)$, until reaching the state of collapse. The simulation was based on 5% mass-proportional damping and restricted to sidesway-only collapse with a drift limit of 7% based on the experimental results. The outcome of this assessment is a structural collapse fragility function, which is a lognormal distribution relating the structure's probability of collapse to the ground-motion intensity, in terms of $S_a(T_1 = 0.6$ s, 5% damping). Figure 4.18 presents the results of nonlinear incremental time-history analyses for QS- and HS-based numerical models.

The mean (the S_a level with 50% probability of collapse) and standard deviation (the dispersion of S_a) for each case can be derived from the following equations (Ibarra and Krawinkler 2005):

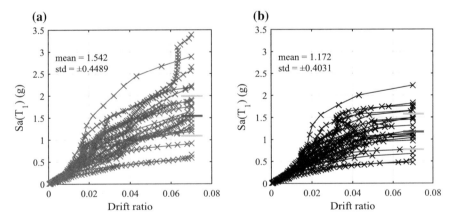

Fig. 4.18 Comparison of IDA results for the RC column: **a** IDA results based on numerical model calibrated to QS test, **b** IDA results based on numerical model calibrated to HS test

$$\ln(\theta) = \frac{1}{n} \sum_{i=1}^{n} \ln(S_a(i)) \qquad (4.1)$$

$$\beta = \sqrt{\frac{1}{n-1} \sum_{i=1}^{n} \left(\ln \left(\frac{S_a(i)}{\theta} \right) \right)^2} \qquad (4.2)$$

where n is the number of ground motions considered and $S_a(i)$ is the S_a value associated with the onset of collapse for the i-th ground motion. In addition, $\ln(\theta)$ and β are, respectively, the mean and the standard deviation of the normal distribution that represents the $\ln(S_a)$ values. Note that the mean of $\ln(S_a)$ corresponds to the median of S_a in the case that S_a is lognormally distributed.

The computed mean and standard deviation values for QS- and HS-based numerical models show that while the dispersion of S_a is similar in both cases, the S_a level with 50% probability of collapse is significantly over-estimated in the QS-based model $(\overline{S_a} = 1.5g)$ compared to the HS-based model $(\overline{S_a} = 1.2g)$. A lognormal cumulative distribution function was then used to define the fragility functions (Porter et al. 2007):

$$P(\text{Collapse}|S_a) = \Phi \left(\frac{\ln \left(\frac{S_a}{\theta} \right)}{\beta} \right) \qquad (4.3)$$

where $P(\text{Collapse}|S_a)$ is the probability that a ground motion with intensity of S_a will cause the structure to collapse, Φ is the standard normal cumulative distribution function (CDF), θ is the median of the fragility function and β is the standard deviation of $\ln(\theta)$.

Fig. 4.19 Comparison of fragility curves for the RC column based on results from QS and HS tests

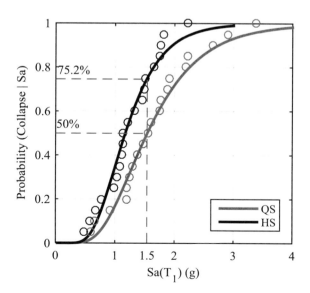

Figure 4.19 indicates that the differences in the collapse probability between fragility curves become larger as the intensity level increases. Specifically, at the intensity level $S_a = 1.5g$, where the QS-based model predicts 50% probability of collapse, the HS-based model predicts 75.2% probability of collapse, which is 1.5 times larger. This shows the significance of the choice of experimental technique and the influence of axial load on collapse risk assessment of the structure.

4.6 Post-Earthquake Rehabilitation of Damaged RC Column Through CFRP Repair

The use of fiber-reinforced polymer (FRP) for the repair and rehabilitation of earthquake-damaged structures can be considered as a cost-effective and time-saving alternative to replacement. This is due to many advantages that FRP possesses over traditional materials for strengthening, including its high tensile strength, light weight, resistance to corrosion, durability and ease of installation. While numerous experimental studies have demonstrated the effectiveness of FRP strengthening for improving the seismic behavior of RC beam–column joints (Shan et al. 2006; Barros et al. 2008; Yalcin et al. 2008; Realfonzo and Napoli 2009; Wei et al. 2009; Gu et al. 2010; Ozcan et al. 2010; ElSouri and Harajli 2011; Realfonzo and Napoli 2012; Wang et al. 2012; Wang and Ellingwood 2015), few studies are available on the repair and rehabilitation of previously damaged column connections and the suitability of FRP as a post-earthquake strengthening solution (Ma and Li 2015; Jiang et al. 2016).

The fourth experiment aimed to evaluate the capability of CFRP repair on rehabilitating the earthquake-damaged RC column to its initial collapse resistance capacity. For this purpose, the damaged column was repaired using CFRP wraps in the plastic zones observed at both ends of the specimen. The damage contained localized zones of spalled and fractured concrete, horizontal and inclined cracking and bent longitudinal reinforcements.

The repair methodology involved: (1) removal of all spalled and fractured concrete; (2) epoxy injection of any cracks wider than 0.3 mm; (3) reinstatement of damaged concrete with a suitable repair mortar; and (4) wrapping of the column with CFRP. However, replacement of the yielded/buckled/ruptured rebars was not included in the repair process. Visual inspection and light tapping using a rubber hammer were used to identify and remove fractured concrete. Cracks that required injection were identified and labeled. Epoxy injection ports were drilled into the concrete directly over the crack and bonded to the surface with epoxy resin. The surface of the crack was sealed and the injection carried out using Sikadur® 52 high-strength adhesive (see Fig. 4.20).

The injection ports were cut after hardening of the Sikadur® 52. A repair mortar, BASF MasterEmaco® S 5300, which is a polymer-modified structural repair mortar, was then used to replace the damaged concrete (see Fig. 4.21). The average

Fig. 4.20 Crack injection application to the upper end of damaged RC column after roughening the concrete surface and rounding all corners

Fig. 4.21 Repair mortar application at the lower (*left*) and upper (*right*) ends of damaged RC column after crack injection

compressive strength of the repair mortar was 41.9 MPa, based on the results of three 50 mm × 50 mm cubes obtained on the test date. The mortar was tested in accordance with ASTM C109 (American Society for Testing and Materials 2011).

The CFRP wrapping was applied over a 600-mm length at each end of the column in regions corresponding to the maximum moment, three days after the crack injection was performed. The concrete in these regions was confined using three layers of MBrace CF130 unidirectional carbon fiber sheet. The CFRP was expected to provide a passive confinement pressure, thereby increasing the compressive strength of concrete. Furthermore, the orientation of the fibers was arranged parallel to the existing steel stirrups, and this was expected to significantly increase the shear capacity at the column ends. The total increases in axial and shear capacity of the column as a result of the CFRP wrapping were estimated as 35 and

Table 4.5 Summary of
CFRP material properties

Properties	MBrace
Tensile strength	4900 (MPa)
Tensile modulus	230 (GPa)
Ultimate elongation	2.1%
Thickness	0.227 mm

Table 4.6 Summary of
epoxy saturant and primer
material properties

Properties	Saturant	Primer
Resin type	Epoxy	Epoxy
Specific gravity	1.12	1.08
Modulus of elasticity	> 3.0 (GPa)	0.7 (GPa)
Tensile strength	> 40 (MPa)	> 12 (MPa)
Compressive strength	> 80 (MPa)	–

250%, respectively, when calculated in accordance with ACI440.2R-08 (American Concrete Institute 2008). A summary of the material properties of the CFRP (MBrace CF130) and adhesive (MasterBrace® P 3500 Primer) used in the repair is given in Tables 4.5 and 4.6.

Prior to the application of the CFRP to the concrete surface, the corners of the column were rounded to achieve a minimum radius of 25 mm. A mechanical abrasion technique was used to remove the weak layer of cement laitance adhering to the surface of the concrete and achieve a surface roughness similar to 60 grit sandpaper (see Fig. 4.22). The surface was cleaned to remove any dust prior to application of the FRP. The FRP was applied using a wet layup technique, and each layer was thoroughly impregnated with resin prior to application to the column. The repair process was performed while the column was still under the MAST system and supporting an axial load of 130 kN, in order to mimic the actual gravity load in a real structural repair scenario. The CFRP was cured at 50 °C for 7 days using heat lamps prior to testing (see Fig. 4.23).

Fig. 4.22 Preparation of RC column surface using mechanical abrasion technique

Fig. 4.23 Application of MasterBrace® P 3500 Primer and MBrace CF130 carbon fiber sheets

4.7 Repeating Mixed-Mode Hybrid Simulation Test

The repaired column was tested under the same hybrid testing conditions as for the initial column. The intensity levels in hybrid simulation included the same previous four scale factors of 0.6, 4.0, 8.0, 9.0, as well as an additional scale factor of 10.0, in order to push the structure to ~0.25% (elastic), 2.0, 4, 6 and 8% maximum inter-story drift ratios, respectively.

Hybrid simulation was completed with no rupture observed in the CFRP sheets. A detailed comparison of hybrid simulation test results for the initial and repaired columns is presented in Fig. 4.24. The results include the hysteretic response in X and Y axes, the axial force time history in Z axis, energy dissipation computed from lateral force deformation in X and Y axes and the biaxial bending moments in Rx and Ry axes.

Figure 4.25a shows a closer view of the hysteretic response of the initial and repaired columns in Y axis, along which the column experienced maximum deformation. Two main significant changes can be observed in the behavior of the repaired column. First, the CFRP repair was not able to restore the flexural strength of the initial column, as the maximum resisting force was 32% less in the repaired column. This is mainly due to the fact that the repair process did not include replacement of the yielded, buckled or ruptured rebars, and as a result the loss of strength could not be fully compensated. Second, the repaired column showed significant improvement in ductility due to the confinement effects of the CFRP wraps. As observed in Fig. 4.25a, the hardening branch of the plastic deformation

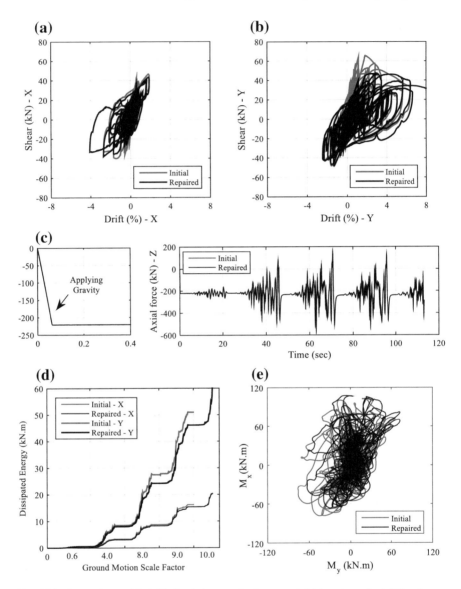

Fig. 4.24 Comparison of hybrid simulation results between initial and repaired RC columns: **a** comparison of lateral force deformation—X axis, **b** comparison of lateral force-deformation—Y axis, **c** comparison of axial load—Z axis, **d** comparison of energy dissipation, **e** biaxial bending moments

response of the repaired column is extended to much larger drifts compared to the initial column. Specifically, while applying the maximum compressive axial load on the initial column (553 kN = 23.35% ultimate capacity), a rapid drop occurred immediately after reaching the peak resisting force. However, the repaired column

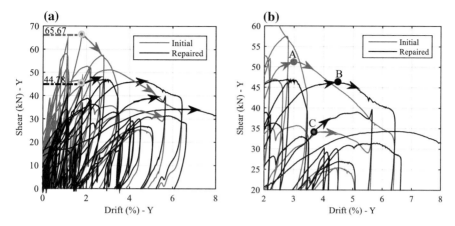

Fig. 4.25 Comparison of the hysteretic response of the initial and repaired columns in Y axis: **a** close view 1, **b** close view 2

remained in the hardening region while being subjected to the same level of axial load. This is also evident by comparisons of other corresponding cycles from the two experiments. For instance, Fig. 4.25b shows, respectively, the capping points 'A' and 'B' for the initial and repaired column from the same corresponding cycles. The initial column shows stiffness hardening up to 3% drift (point 'A'), while this value has been extended to 4.5% drift (point 'B') for the repaired column. In addition, by comparing the behavior of the RC columns after point 'C,' which is located on the same corresponding cycle and at the same level of drift, it is observed that the initial column entered the post-capping negative stiffness region, while the repaired column was still in the stiffness hardening region.

4.8 Comparison of Initial and CFRP-Repaired RC Columns

4.8.1 Numerical Model and Calibration

The same numerical model, as shown in Fig. 4.13, was selected for the comparative collapse risk assessment of the initial and repaired columns. Figure 4.26 compares the calibrated SDOF model of the column to the hybrid simulation results in Y axis. Particular emphasis was placed on precisely mimicking the plastic and post-capping deformation capacities, which are known to have an important influence on collapse prediction. Table 4.7 compares the IMK model parameters for the calibrated models. The main differences include the use of a lower flexural strength but a much higher plastic deformation capacity in the stiffness hardening region for the model of the repaired column.

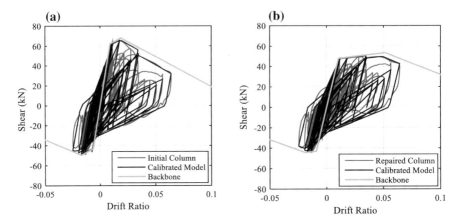

Fig. 4.26 Calibration of SDOF model to hybrid test results: **a** calibration of initial column response, **b** calibration of repaired column response

Table 4.7 Comparison of IMK model parameters calibrated to experimental results

Experiment method	M_c/M_y	M_y^+ (kN/m)	M_y^- (kN/m)	EI/EI_g	ψ_p^+	ψ_p^-	ψ_{pc}^+	ψ_{pc}^-	λ_S, λ_C
Initial column	1.12	78.0	54.0	0.258	0.07	0.12	0.9	1.0	9.2
Repaired column	1.10	56.0	54.0	0.258	0.34	0.12	0.9	1.0	9.2

4.8.2 Fragility Analysis

Incremental dynamic analyses (IDAs) were performed using the calibrated numerical models in order to capture a range of probable dynamic response behaviors due to record-to-record variability in ground-motion characteristics. The ground motions used for this purpose are the same as those previously listed in Table 4.4. Each unidirectional ground motion was individually applied to the calibrated SDOF models and increasingly scaled, until the state of complete collapse. The simulation was based on 5% mass-proportional damping and restricted to sidesway-only collapse with a drift limit of 7%, based on the experimental results. The outcome of this assessment is a structural collapse fragility function for the initial and repaired columns, respectively, which is a lognormal distribution relating the structure's probability of collapse to the ground-motion intensity level (S_a). Figures 4.27 and 4.26 present the results of nonlinear incremental time-history analyses and the associated fragility curves, respectively. It is observed that at an intensity level (S_a) of 1.2 g, the probability of collapse for the initial columns is 50%, while this value for the repaired column is 44%. This shows that CFRP repair can effectively restore the capacity of the column and slightly improve the resistance of the column to sidesway collapse (Fig. 4.28).

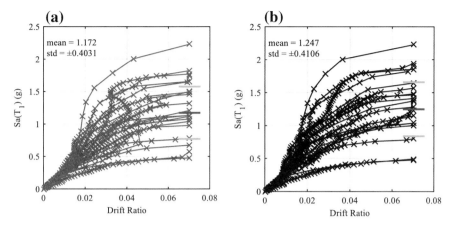

Fig. 4.27 Comparison of IDA results for the SDOF models: **a** IDA results based on the model calibrated to initial RC column response, **b** IDA results based on the model calibrated to repaired RC column response

Fig. 4.28 Comparison of fragility curves for the initial and repaired RC columns

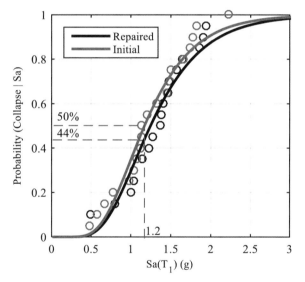

References

American Concrete Institute. (2008). *Guide for the design and construction of externally bonded FRP systems for strengthening concrete structures.* SN:440.2R-8, Farmington Hills, U.S.

American Society for Testing and Materials. (2011). *Standard test method for compressive strength of hydraulic cement mortars (using 2-in. or [50-mm] cube specimens).* SN:C109–11/C109 M-11, West Conshohocken, U.S.

Barros, J. A. O., Varma, R. K., Sena-Cruz, J. M., & Azevedo, A. F. M. (2008). Near surface mounted CFRP strips for the flexural strengthening of RC columns: Experimental and numerical research. *Engineering Structures, 30*(12), 3412–3425.

Deierlein, G. G., Reinhorn, A. M., & Willford, M. R. (2010). *Nonlinear structural analysis for seismic design, a guide for practicing engineers*. Gaithersburg, MD, U.S.: National Institute of Standards and Technology.

Del Carpio Ramos, M., Mosqueda, G., & Hashemi, M. J. (2015). Large-scale hybrid simulation of a steel moment frame building structure through collapse. *Journal of Structural Engineering, 142*(1), 04015086.

ElSouri, A. M., & Harajli, M. H. (2011). Seismic repair and strengthening of lap splices in RC columns: Carbon fiber-reinforced polymer versus steel confinement. *Journal of Composites for Construction, 15*(5), 721–731.

Federal Emergency Management Agency. (2007). *Interim testing protocols for determining the seismic performance characteristics of structural and nonstructural components*. Department of Homeland Security (DHS), Washington D.C., U.S.

Gu, D. S., Wu, G., Wu, Z. S., & Wu, Y. F. (2010). Confinement effectiveness of FRP in retrofitting circular concrete columns under simulated seismic load. *Journal of Composites for Construction, 14*(5), 531–540.

Haselton, C. B., Liel, A. B., Lange, S. T., & Deierlein, G. G. (2008). *Beam-column element model calibrated for predicting flexural response leading to global collapse of RC frame buildings*. Berkeley, U.S.: Pacific Earthquake Engineering Research Center, University of California.

Hashemi, M. J., Masroor, A., & Mosqueda, G. (2014). Implementation of online model updating in hybrid simulation. *Earthquake Engineering and Structural Dynamics, 43*(3), 395–412.

Hashemi, M. J., & Mosqueda, G. (2014). Innovative substructuring technique for hybrid simulation of multistory buildings through collapse. *Earthquake Engineering and Structural Dynamics, 43*(14), 2059–2074.

Hashemi, M. J., Mosqueda, G., Lignos, D. G., Medina, R. A., & Miranda, E. (2016). Assessment of numerical and experimental errors in hybrid simulation of framed structural systems through collapse. *Journal of Earthquake Engineering, 20*(6), 885–909.

Ibarra, L. F., & Krawinkler, H. (2005). *Global collapse of frame structures under seismic excitations*. Stanford, U.S.: Stanford University, John A. Blume Earthquake Engineering Center.

Ibarra, L. F., Medina, R. A., & Krawinkler, H. (2005). Hysteretic models that incorporate strength and stiffness deterioration. *Earthquake Engineering and Structural Dynamics, 34*(12), 1489–1511.

Jiang, S.-F., Zeng, X., Shen, S., & Xu, X. (2016). Experimental studies on the seismic behavior of earthquake-damaged circular bridge columns repaired by using combination of near-surface-mounted BFRP bars with external BFRP sheets jacketing. *Engineering Structures, 106*, 317–331.

Lynn, A. C., Moehle, J. P., Mahin, S. A., & Holmes, W. T. (1996). Seismic evaluation of existing reinforced concrete building columns. *Earthquake Spectra, 12*(4), 715–739.

Ma, G., & Li, H. (2015). Experimental study of the seismic behavior of predamaged reinforced-concrete columns retrofitted with basalt fiber–reinforced polymer. *Journal of Composites for Construction*. doi:10.1061/(ASCE)CC.1943-5614.0000572.

Nakamura, T., & Yoshimura, M. (2002). Gravity load collapse of reinforced concrete columns with brittle failure modes. *Journal of Asian Architecture and Building Engineering, 1*(1), 21–27.

Nakata, N. (2007). *Multi-dimensional mixed-mode hybrid simulation, control and applications*. Ph.D. Dissertation. U.S.: University of Illinois at Urbana-Champaign.

Ozcan, O., Binici, B., & Ozcebe, G. (2010). Seismic strengthening of rectangular reinforced concrete columns using fiber reinforced polymers. *Engineering Structures, 32*(4), 964–973.

Pan, P., Nakashima, M., & Tomofuji, H. (2005). Online test using displacement-force mixed control. *Earthquake Engineering and Structural Dynamics, 34*(8), 869–888.

PEER. (2013). *Structural performance database*. Berkeley, U.S.: Pacific Earthquake Engineering Research Center, University of California.

Porter, K., Kennedy, R., & Bachman, R. (2007). Creating fragility functions for performance-based earthquake engineering. *Earthquake Spectra, 23*(2), 471–489.

Realfonzo, R., & Napoli, A. (2009). Cyclic behavior of RC columns strengthened by FRP and steel devices. *Journal of Structural Engineering, 135*(10), 1164–1176.

Realfonzo, R., & Napoli, A. (2012). Results from cyclic tests on high aspect ratio RC columns strengthened with FRP systems. *Construction and Building Materials, 37,* 606–620.

Schellenberg, A. H., Mahin, S. A., & Fenves, G. L. (2009). *Advanced implementation of hybrid simulation*. Berkeley, U.S.: Pacific Earthquake Engineering Research Center, University of California.

Scott, M. H., & Fenves, G. L. (2006). Plastic hinge integration methods for force-based beam-column elements. *Journal of Structural Engineering, 132*(2), 244–252.

Sezen, H. (2002). *Seismic response and modeling of reinforced concrete building columns*. Ph.D. Dissertation. Berkeley, U.S.: University of California.

Shan, B., Xiao, Y., & Guo, Y. R. (2006). Residual performance of FRP-retrofitted RC columns after being subjected to cyclic loading damage. *Journal of Composites for Construction, 10*(4), 304–312.

Wang, N. Y., & Ellingwood, B. R. (2015). Limit state design criteria for FRP strengthening of RC bridge components. *Structural Safety, 56,* 1–8.

Wang, T., Mosqueda, G., Jacobsen, A., & Cortes-Delgado, M. (2012). Performance evaluation of a distributed hybrid test framework to reproduce the collapse behavior of a structure. *Earthquake Engineering & Structural Dynamics, 41*(2), 295–313.

Wei, H., Wu, Z. M., Guo, X., & Yi, F. M. (2009). Experimental study on partially deteriorated strength concrete columns confined with CFRP. *Engineering Structures, 31*(10), 2495–2505.

Wibowo, A., Wilson, J. L., Lam, N. T. K., & Gad, E. F. (2014). Drift performance of lightly reinforced concrete columns. *Engineering Structures, 59,* 522–535.

Wang, Z. Y., Wang, D. Y., Smith, S. T., & Lu, D. G. (2012). Experimental testing and analytical modeling of CFRP-confined large circular RC columns subjected to cyclic axial compression. *Engineering Structures, 40,* 64–74.

Yalcin, C., Kaya, O., & Sinangil, M. (2008). Seismic retrofitting of R/C columns having plain rebars using CFRP sheets for improved strength and ductility. *Construction and Building Materials, 22*(3), 295–307.

Zhong, W. (2005). *Fast hybrid test system for substructure evaluation*. Ph.D. Dissertation. U.S.: University of Colorado Boulder.

Chapter 5
Closure

Abstract This chapter presents a summary of key contributions and concluding remarks. Research areas for further development and study are also briefly discussed.

Keywords MAST system · Collapse assessment

5.1 Summary and Conclusions

Over the past two decades, hybrid simulation has received increasing attention from the earthquake engineering community, as it provides a platform to accurately simulate the nonlinear dynamic response of large-/full-scale structural systems with reduced cost and efforts. In this method, only sensitive and critical components of a structural assembly that is difficult to model numerically are substructured in the laboratory, and the remainder of the assembly, inertia and damping forces and gravity and dynamic loads are simulated using finite-element analysis software.

The quality of the hybrid simulation test strongly depends on the correct application of the interface boundary conditions between the numerical and the physical sub-domains. This, however, poses a major challenge, as the actions on structures during extreme events such as earthquakes are generally multi-directional and continuously varying, and due to laboratory limitations, the simulation of such highly coupled multi-directional loading conditions can be difficult to achieve.

The Multi-Axis Sub-structure Testing (MAST) system at Swinburne University of Technology has been developed to expand the capabilities of large-scale hybrid simulation tests by allowing the experimental simulation of complex time-varying 6-DOF boundary effects using mixed load/deformation modes. The hydraulic

---

The original version of this chapter was revised: See the "Chapter Note" section at the end of this chapter for details. The erratum to this chapter is available at https://doi.org/10.1007/978-981-10-5867-7_6

© The Author(s) 2018   73
R. Al-Mahaidi et al., *Multi-axis Substructure Testing System for Hybrid Simulation*,
SpringerBriefs in Structural Mechanics, https://doi.org/10.1007/978-981-10-5867-7_5

actuator/control system of the MAST consists of four $\pm$ 1 MN vertical actuators, two pairs of $\pm$ 500 kN horizontal actuators in orthogonal directions and an advanced servo-hydraulic control system capable of imposing simultaneous 6-DOF states of deformation and load in switched- and mixed-mode control. The reaction system of the MAST consists of a 9.5-tonne steel cruciform crosshead that transfers the 6-DOF forces from the actuators to the specimen and an L-shaped strong wall (5 m tall $\times$ 1 m thick) and a 1-m-thick strong floor. The hybrid simulation framework of the MAST uses an advanced three-loop hybrid simulation architecture including the innermost servo-valve control loop that contains MTS Flex Test Controller, the middle actuator command generation loop that runs on the xPC-Target real-time digital signal processor (DSP) and includes the predictor-corrector algorithm and the outer integrator loop that runs on the xPC-Host and includes OpenSees, OpenFresco and MATLAB/Simulink.

The MAST system was used in application for two series of experiments. The objective of the first series was to compare the application of conventional quasi-static (QS) versus hybrid simulation (HS) test results on the estimation of the collapse probability of RC structures. Two large-scale experimental tests were conducted on identical large-scale limited-ductile RC columns by the respective testing methods using the MAST system. Larger flexural strength and significant reduction in drift capacity resulted from the HS test due to the higher levels of axial loads. In addition, lower levels of cyclic degradation were observed in the HS test due to the ratcheting of the structure's lateral deformation, whereas the specimen experienced large cycles and load reversals before failure in the QS test. The hysteretic response behaviors from the QS and HS tests were then used, respectively, for calibrating the numerical models, which were employed for comparative collapse risk assessment. The fragility curves obtained show significant deviation in the estimated collapse probability, which may have crucial implications for seismic design and risk assessment.

The objective of the second series of experiments was to use hybrid simulation to evaluate the effectiveness of CFRP repair on restoring the resistance capacity of earthquake-damaged RC structures against collapse. For this purpose, the damaged specimen from the previous hybrid simulation was repaired using CFRP wraps and retested under the same loading conditions. From the comparison of experimental results, significant enhancement of ductility was observed for the repaired column, while the strength was not fully recovered as the yielded, buckled or ruptured rebars of the damaged column were not replaced in the repair process. A comparative collapse risk assessment of the initial and repaired RC columns was performed using the experimental results. The fragility curves obtained from these simulations show that the collapse risk of the CFRP-repaired column is slightly lower than that of the initial column, highlighting the suitability of using CFRP as a post-earthquake strengthening solution.

5.2 Future Work

For many years, one of the major challenges in collapse assessment of structures has been the lack of realistic and reliable data from structures collapsing and confidence in the prediction of a structure's response as it approaches collapse. This has been mainly due to the complexity of the physical interactions of structural elements at imminent collapse and difficulties in capturing these complexities in a cost-effective and reliable manner.

The unique and versatile capabilities of the MAST system provide a powerful tool for structural and earthquake engineers to investigate the dynamic behavior of large-scale structures and characterize their safety margins beyond the design level, all the way to the state of complete collapse. This plays an important role in decision making in terms of protecting the safety of users and managing economic losses, depending on different possible outcomes of extreme events.

Erratum to: Multi-axis Substructure Testing System for Hybrid Simulation

Riadh Al-Mahaidi, Javad Hashemi, Robin Kalfat, Graeme Burnett and John Wilson

Erratum to:
R. Al-Mahaidi et al., *Multi-axis Substructure Testing System for Hybrid Simulation*, **SpringerBriefs in Structural Mechanics, https://doi.org/10.1007/978-981-10-5867-7**

In the original version of the book, the following corrections were carried out:

Open Access was removed, including Open Access logos on the cover and in the Frontmatter. The copyright information in the Frontmatter and at chapter-level was also corrected.

The book has been updated with the changes.

The updated original online version of this book can be found at
https://doi.org/10.1007/978-981-10-5867-7

Appendix

The details and specifications of the cruciform crosshead and the hydraulic actuator dimensions are presented in this appendix (Figs. A.1, A.2, A.3 and A.4).

© The Author(s) 2018
R. Al-Mahaidi et al., *Multi-axis Substructure Testing System for Hybrid Simulation*,
SpringerBriefs in Structural Mechanics, https://doi.org/10.1007/978-981-10-5867-7

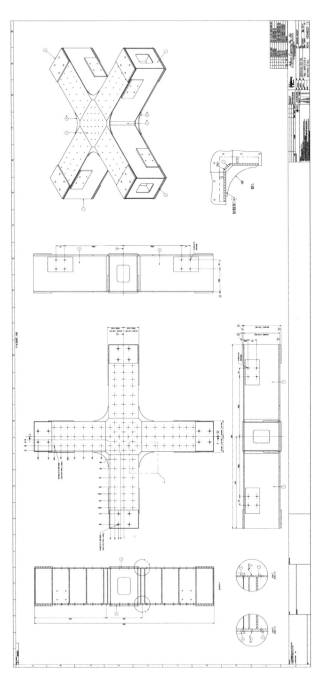

Fig. A.1 Cruciform overall details

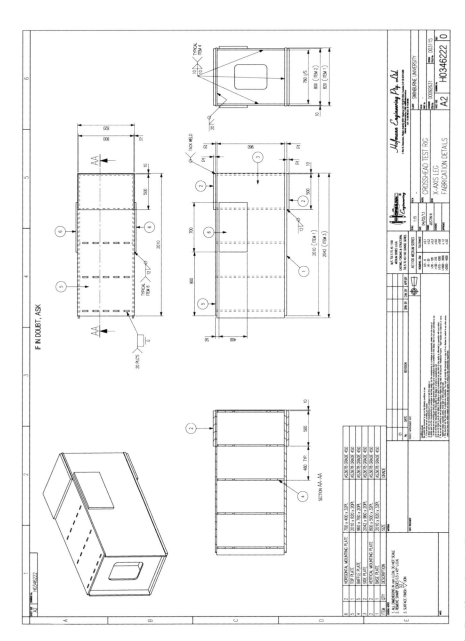

Fig. A.2 Cruciform leg details

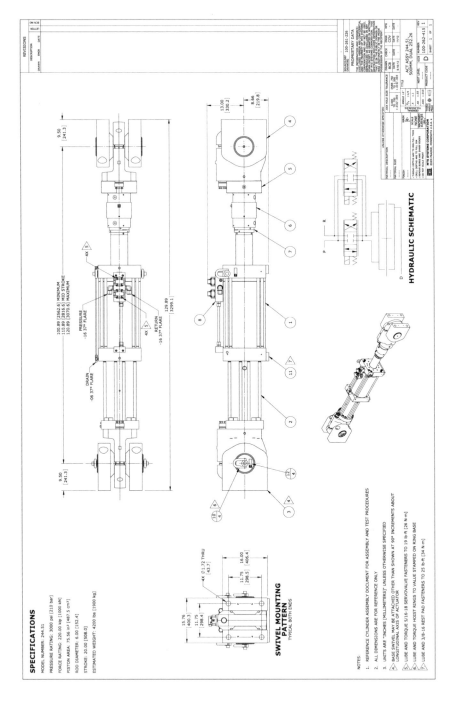

Fig. A.3 Vertical actuator specifications

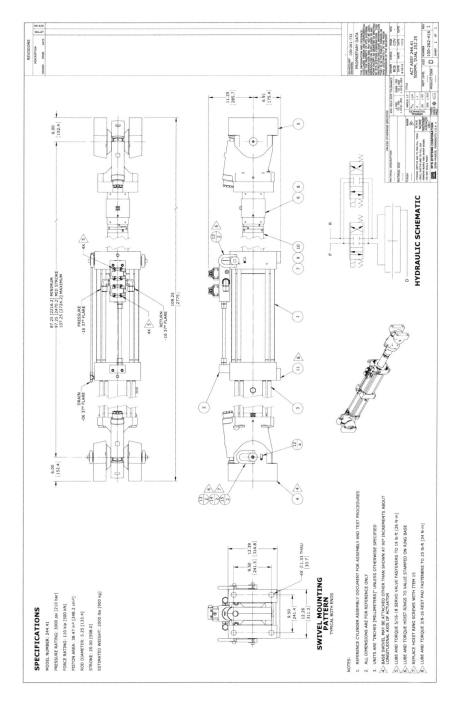

Fig. A.4 Horizontal actuator specifications

Printed in the United States
By Bookmasters